THE MODERN ANCIENTS

L. Mason Jones served a number of years in the military, and travelling on a so-called "government service" passport, found himself in such places as south Yemen, Bahrain, the Gulf of Oman, Cyprus and Germany. After leaving the service, he became part of the team producing the highly successful business jet, The Hawker 125. He functioned as a quality-engineering inspector with, initially, British Aerospace then Corporate Jets Inc and finally Raytheon USA, the latter of which purchased the thriving business and moved production to the USA. Mr Jones then left the business to concentrate on writing projects. He has three adult offspring and resides in Chester.

THE MODERN ANCIENTS

L. MASON JONES

Arena Books

Copyright © L. Mason Jones 2022

The right of L. Mason Jones to be identified as the author of this work has been asserted in accordance with the Copyright, Design and Patents Act 1988.

First published by Arena Books in 2022

Arena Books
6 Southgate Green
Bury St. Edmunds
IP33 2BL.

www.arenabooks.co.uk

All rights reserved. Except for the quotation of short passages for the purposes of criticism and review, no part of this publication may be reproduced, stored in a retrieval system, or transmitted, in any form or by any means, electronic, mechanical, photocopying, or recording or otherwise, without prior permission of the publisher.

L. Mason Jones

The Modern Ancients

British Library cataloguing in Publication Data. A Catalogue record for this book is available from the British Library.

ISBN 978-1-914390-09-8

BIC categories: PDA, PGK, RBC, RBGD, PDZ, ABGE, PGS, RNR, TTD.

Distributed in America by *Ingram International, One Ingram Blvd., P.O. Box 3006, La Vergne, TN 37086-1985, USA.*

Cover design by Jason Anscomb

I would like to dedicate the publication of this book to the memory of the late Andrew Tomas of Australia, who travelled widely and studied so many ancient documents and accounts. He brought to light many riddles of ancient science and recorded them diligently in a book published fifty years ago. As Solomon said, "Can anyone today say, 'see this, it is new?' It hath been done by people before us, of old time," and so this author is happy to stand on the great shoulders of one who went before.

CONTENTS

INTRODUCTION		1
CHAPTER I	HOW OLD IS TECHNOLOGY?	6
CHAPTER II	THE FACTORY ACT 2,500 B.C.	41
CHAPTER III	THE ANCIENT MINERS	47
CHAPTER IV	THE ANCIENT DOCTORS	51
CHAPTER V	THE ARK OF MOSES	60
CHAPTER VI	MAKE ME SOME GOLD	68
CHAPTER VII	ODDITIES OF EARTH AND COSMOS	93
CHAPTER VIII	ANCIENT SCIENCE	128
CHAPTER IX	NOAH AND THE WORLD SURVEYORS	146

INTRODUCTION

As we today, in the twenty-first century, bask in our technological, medical and scientific advancements, we must ask, are we really the first? It is becoming clear in the revelations emerging from the study of ancient written works and of ancient sacred sites, that many of the things we think we are discovering for the first time today, may well have been discovered before.

It is strange to relate that King Solomon was already making the same observation nearly three thousand years ago. The revelations emerging in our own time from the study of archaic books and documents seems to imply that many scientific ideas and notions that we accept today, not only occurred during the time of the people of antiquity, but were also put into practice by them. Moreover, what appear to us to be scientifically advanced ideas, seem to have come upon the people of the past almost as though they came from an outside source or from unknown "teachers". And it is certainly becoming increasingly evident that we have vastly underestimated the technical skills, ability and knowledge of people that existed in pre-history.

Just as occurs today, in our own history, new ideas were perhaps too soon rejected before they had been thoroughly researched. However, whereas today one might say "OK fair enough, you are entitled to your opinion," in times of old, there may have been more serious consequences depending on the subject that generated the idea. Enlightened people of older times may have been at the mercy of more narrow-minded rulers and therefore often paid with their lives for announcements of a cosmic nature that may have contravened the religious dogma of the time.

One must admire the courage of their convictions, knowing how dangerous it was to announce unpopular ideas. Of all the sciences, especially during the medieval period and early-modern periods, the study of the heavens was the most risky to be involved in; and if a learned man made a significant discovery, it would have been difficult to keep quiet about it despite the danger.

We mentioned King Solomon, it is clear that he was well educated in the discoveries that went before him. In *Pillars of Fire*, I quoted Solomon's remarks, "Can it be said today, of things discovered, 'see this, it is new. It hath been done long before us of old time?' we owe a great debt to those who went before us and should not be afraid to pay it back when venturing into realms that we may not have dared to tread before." This is particularly relevant with regard to our own new understandings of the Universe and so we ask the question, "Could all this wonderment have been chance, based on the idiosyncrasies of the impartial science?" Such learned men as Galileo Galilei, Giordano Bruno and Charles Darwin compelled us to think more deeply about our surroundings, our world, the vast universe, the implications of creation and our very existence.

Yet although these learned people of science, geology and archaeology contributed to the idea that our world is much more ancient than we had previously thought, even they could not get it right all of the time. The strange thing is, is that many of the ancient Greek philosophers, in particular some of their pronouncements on celestial matters, seem to have been closer to the truth (than our more recent scientists) all along. And the later astronomers we have mentioned began to move more toward the line of thinking of their ancient counterparts as time progressed.

Scientists, whose revelations caused any kind of shock to the establishment, whether that establishment was religious or "Pagan", paid with their lives. Giordano Bruno was one such unfortunate, and Charles Darwin escaped the flames, only because the punishments were far less severe than those meted out in the centuries before him. But there is no doubt that there were some people amongst the ecclesiastical fraternity that wished upon Darwin the same fate as Bruno; such was the shock and horror that Darwin's arguments about human evolution evoked in them.

Introduction

However, it was not just the leaders of the Church who were affected by Darwin's ideas; thousands of people all over the world were made to rethink the whole idea of how they came to be. And Charles Darwin's scope was wide; he gave us much more than a new theory of human development – he also travelled the world and made new discoveries in the study of botany and flora and fauna. He could even be referred to as (as I have done in other work) the "New Messiah" which, in a sense, he was, having so dramatically affected the lives and beliefs of so many people.

But strangely, despite the knowledge and discoveries of these great individuals in our own recorded history – it appears that over a wider span of time, the advancement of knowledge and technology has not appeared as a steady and continuous line on a graph. If it had done so, without dramatic interruptions of some kind – including war, disease, persecution and the religious interference that prevailed after the Egyptians and ancient Greeks – perhaps we would be looking up today at a planet that was more blue than red. That is, we might be getting on with the business of tilling our second home, Mars, with the sweat of our brows as we pause to look at our original homeland of the Earth. Instead, with this erratic line of advancement, it is clear that much scientific knowledge and development has been seriously hampered by fear and ignorance.

The blame for this must be laid largely at the doors of the church in the Middle Ages. Religion has no doubt brought succour and comfort to many and indeed still does, and a much more enlightened approach exists today with regard to revelations, such as Darwin's, that were once thought of as satanic. Religion has gradually been accepting and incorporating new scientific revelations in the same way that the practice of burning alleged witches was finally outlawed. And if the church could survive the trauma and shock of being told that apes, rather than God, were responsible for our human appearance, then it could survive anything that science could throw at it. Nevertheless, this mellowing did not necessarily occur in all quarters.

As noted, there have been periodic crescendos and diminuendos in human history and strangely, even during that time of great advancement that was the Victorian Age, outmoded beliefs persisted. During this time, when it was said that the Sun never set on the British Empire (and when children were shown the vast swathes of red on maps that depicted British rule around the globe), many people, due to allegedly logical calculations by senior church members such as James Usher the Archbishop of Armagh, still accepted that the world was created in 4004 B.C. These calculations were based on the calculated ages of the patriarchs. And in turn, this meant accepting the incredible ages of the patriarchs, some of whom were said to have lived for almost 1000 years. But, as well as the scientists, industrialists, designers, bridge builders and the maritime engineers, of the nineteenth and twentieth centuries, there were other "enlightened ones" who were destined to provide other shocks to the church albeit lesser ones than Darwin's Origin of Species had caused.

Together, a new breed of geologists and archaeologists eventually put a stop to the former outmoded forms of thinking which they viewed as preposterous. Among them was the French Naturalist Comte de Buffon who, although widely far off the mark, was heading in the right direction when he proposed that the Earth cooled down some 35,000 years ago and that life appeared some 20,000 years later, around 15,000 B.C. These calculations were certainly more rational than the good Archbishop's earlier calculations, but the estimated time of the event was still in full retreat. All of these individuals were quite convinced of the accuracy of their calculations.

In 1862, Lord Kelvin added a further ten million years to the French naturalist's calculations (or estimations) regarding the true age of the Earth. And now, we assume that we have got it right and that the true age of the Earth is 4 ½ billion years old, but are we to assume that the matter

Introduction

is settled? It has been suggested that the Earth is much older still. So the estimated age of the Earth has jumped from Buffon's 35,000 years to 4 ½ billion years, and who is to say it will not be pushed back even further? It remains to be seen whether our current estimates stay securely in place with current or future inventions of more advanced and sophisticated instruments, and methods of dating organic and mineral matter. We may yet have to revise our current views and estimates. This is already apparent with some indications arising from analysis of the Baltic crust that seem to indicate that 7 billion years is the true age of the Earth. This means simply "hold the front page." After all, the same thing has happened with regard to the estimates of the initial creation, that is, the so- called "Big Bang".

CHAPTER I
HOW OLD IS TECHNOLOGY?

If we, for a moment, take the aforementioned Bishop Usher's line of thinking – we will attempt to compute the date of birth of the first man, Adam (according to Genesis), via the ages of the patriarchs (who were the descendants of Adam). Moses, who wrote Genesis, seemed very sure of the ages of the patriarchs because he gave very precise dates of their births and deaths. He was also quite specific with regard to their qualifications, trades and skills – which makes this information seem more likely to contain truth – and so herein lies a problem. If we take Moses' dates as accurate, and analyse Adam's birth date via the great ages of his descendants, then his date of birth can be calculated as being at some point between 4,000 and 5,000 B.C.

But if this birth date of the first man is anywhere near accurate, and if, as archaeology is increasingly showing, technology was more far advanced in the two or three millennia years before the birth of Jesus than we have hitherto accepted, then where did this relatively instant learning, knowledge and technology come from? We would have to accept that technology and culture arose at an extremely rapid rate in the years after Adam's birth, for it to have been as advanced as it was by around 2,500 B.C. Adam, being the first, would have had to learn everything from scratch. The story in Genesis says that he was instructed to work hard to till the land, but if the timeline is correct, then he must have had significant help from the creator, or should we say creators?

In the Genesis list, we find "Cainan" described as a craftsman. His equivalent in the Chaldean list was "Ammenon", which also means "craftsman". The Bible, in particular the Old Testament, has had its fair share of criticism, as indeed has the New Testament, with regard to the working of "miracles" which can be mostly be explained away today as magic tricks. But it is Moses who has been the focus of most of the primary points of criticism, in particular, his story of human creation and its timeline. Yet despite the criticism, and despite analysis by sceptical science, it remains firmly entrenched in many people's minds as a truthful account. This determination to hold on to the truth of the story is strange in itself considering the post-enlightenment world that we live in and our current level of technological advancement.

Nevertheless, it is obvious to anyone who scrutinises the Bible text closely that it has many anomalies – for example in its tendency to be vague (the birth of Adam) at one point, and quite specific in the next. One such anomaly lies in the immense ages of the patriarchs. If this is true, then they indeed may have achieved great levels advancement in their knowledge and skill – but why did Moses not explain, in Genesis, the discrepancy between the ages of the patriarchs, and the ages of people in his own time whose lifespans were closer to our own?

Enoch is a character who appears in the Hebrew (not the Christian) bible, as well as the Dead Sea Scrolls, as a patriarch who lived before the flood. He must have been quite a special being and may have been in possession of extremely enlightening information regarding the cosmos, (nearer my Gods to thee). He was described as "a bearer of divine revelations" and "he, to whom the secrets of Heaven and Earth are revealed." Could it be that when he spoke of his "revelations", he talked himself out of the Christian Old Testament? Today, while our astronomers are busily searching for the secrets of "heaven" (i.e. the Universe) perhaps they are looking for secrets that Enoch already discovered thousands of years ago. And the S.E.T.I (Search for Extra-Terrestrial Intelligence) community would certainly do well to study closely any more scrolls that emerge regarding this character.

Fortunately for Enoch, during his time, he was more revered than feared and did not meet the same fate as the much later Dominican Monk Giordano Bruno, who (as we have said in *Pillars of Fire*) was burnt alive at the Piazza De-Fiore in Rome as a Heretic. Today, Bruno, with his enlightened and advanced knowledge of all things celestial, would no doubt have been a useful member of the S.E.T.I team.

THE MODERN ANCIENTS

In one of his writings, Giordano Bruno stated that there are an infinite number of suns and that there are planets that revolve around them. Moreover, he added that some of these worlds may be populated. So as a possessor of such enlightened knowledge, who might Bruno's tutor have been? He could surely not have guessed at this accurate information by himself. And, even if we could answer the question regarding Bruno's tutor, we would have to ask the same question regarding that person. It would have to stop somewhere, and I suggest here that there must have been a common teacher, or teachers, back in history at some point. When we mentioned Enoch being in possession of such "enlightening" information, we must ask, who among the other patriarchs was as equally enlightened? All this technology appears to have come to them somewhat instantaneously, and this also applies to their "arts and crafts."

Certainly, Noah was a craftsman or he could not have achieved all that he did (as I related in *Strange Realities*) with regard to his high-tech marine design skills, navigational skills, organisational abilities, and in particular, scouring the world for all the species of animals (in pairs) some of which were not available in his own land, (we could mention such creatures as arctic foxes, polar bears and so-forth). And with regard to Enoch, his name also means "devoted", one who is initiated into "secret learning", and "teacher". So, what were these secrets, and who "initiated" him? In the Old Testament, Lamech is shown as the father of Noah, and he lived during the period when the arts emerged. Everything seems to indicate a higher technology and advanced learning that came upon them all, quite suddenly.

The first wife of Lamech, Adah, was described as an artist. There is nothing surprising in this when we reflect on the amazing feats of art depicted in the Magdalenian cave art, which incidentally, was completed even before Moses' "first man Adam" – in fact some 35,000 years before him. Raising once again the anomalies of the Old Testament and Moses' account of Adam and creation. The second wife of Lamech was called Zillah, and she was a maker of sound, a "player," a point that surely

How Old is Technology?

indicates the ancientness of music. The second son of Adah, the artist, was called Jubal and he is described as a "teacher" or father of all of such things as the harp, the organ and the stringed instruments. Such things did not come into existence overnight, they are quite technically constructed with special types of wood and so-forth, not to mention the special strings important to the sound. It is hard to imagine an ancient churchman piously playing his organ in a chapel just a few hundred years after Adam.

Nevertheless, Genesis maintains that as long ago as 3,860 B.C., technology made from metals existed. Genesis 4:22 refers to a patriarch called Tubal-Cain, who was the son of the aforementioned Zillah and he was obviously an experienced artificer and metal fabricator. He was an instructor of all artificers who worked with brass and iron. Brass is an alloy; therefore, he must have been experienced in the alloying of metals, which would of course include steel in its various grades, alloyed from iron.

In Chapter 2, we will discuss the amazing archaeological discovery near the mountains of Ararat, and it will be shown that, as ancient (yet modern) as they were, we have evidence that Tubal-Cain was indeed teaching on the subject as early as 3860 B.C. His teachings became a firm foundation as seen in the fact that already in 2500 B.C. foundries existed. Evidently, to do all this fabricating and alloying of metals, foundries must have already been set up and running during the time of Tubal-Cain.

Clearly, with all this metal technology occurring, there must also already have been various types of machinery in existence at this time, for example, machinery for bending, cutting, and fabricating metals, and various types of presses to form the metals into brackets and structures. We must ask how were they powered, some must have required hydraulic power. As well as hydraulics, today, for such operations, we need to employ pneumatic power (requiring compressors), electric power also requiring separate generators, or the ability to tap into the National Grid via plugs and sockets. Granted, some of these machines could have been step operated or have relied on leverage as in some metal cutting machines, but certainly not all of them.

We know of course that copper was being mined so long ago that even peoples whose history goes back many thousands of years cannot explain the origins of its use. Tin, zinc, copper and iron were all already obviously used and in existence during the time of the aforementioned Tubal-Cain. Even the phrase "Tubal-Cain" is said to mean "the 'brass' of Cain." If Cain was aware of the alloy brass, he has come a very long way in a very short time from his father and his earlier pre-occupation with tilling the Earth with the sweat of his brow.

And so we return to the idea of "ancient tutors". Perhaps these tutors were the same people who were responsible for the very existence of Cain's father Adam; perhaps they were a breed of people who, when they came to Earth, decided to "make" men in their image. With this line of reasoning, after having given this advanced creative intelligence and technical knowledge to humanity, it follows that they would be anxious for humans to advance as rapidly as possible. And perhaps one of the ways they did this was to point out which areas to excavate for all the necessary ingredients required for smelting, fabricating and forging, and by teaching them about the alloying processes.

I have suggested in another work that this was the first giant leap for mankind, the second being the fashioning of metals into a spacecraft and the third being that which happened in July 1969 with Apollo 11. There is no question that the progression of technology from Adam to blast furnaces was comparable to the astounding advancement that our era experienced in the form of the great leap that the Wright brothers achieved in aero-technology. Yet, once again, we will show that even with regard to flying craft, we may not have been the first – the ancients could well have beaten us to it in this technology also.

However, we have to ask whether technological know-how automatically goes hand-in-hand with human wisdom; even in such advanced times, some of the various tribes that had established themselves after the great surveying and mapping of the world (an event that we will explore in the coming chapters) would also have manufactured weapons as one of their metallurgical products. War and conflict have been in existence ever since mankind first appeared on Earth and no doubt will continue to do so unless our advancements in genetic science finally put a stop to it by influencing mental advancement to the point of eliminating the genes that cause humans to abdicate completely from rational behaviour.

The fact that the ancients did include metal implements in war, is made clear by the mention in Genesis of Methuselah, the eighth Patriarch, who is described as a "man of the arrow" and "the flying dart." Here we have an obvious distinction between the simple arrow and this missile of some type. "Flying darts" are mentioned in other ancient texts, implying a more fearsome technology than a simple arrow fired from a medieval bow. The wars and conflicts depicted in the Hindu epics seem to hint at a knowledge of weaponry that seems nuclear in its description. The description of a "flying dart" in the Hindu sagas and in the ancient book of the *Mahabharata* and its fearsome stories, seems to describe something like a nuclear missile that was used by one side against their enemies.

While written accounts can be disregarded as exaggerated tales, earthly evidence cannot. The Hindu sagas relay the use of what sound like highly destructive weapons that seem nuclear in their capability. We will explore the physical evidence for this later. A raging eighteen-day war is described that is said to have occurred between a group of people known as the Kauravas and another group that were called the Pandavas who according to the texts occupied the upper regions of the River Ganges. As well as this conflict, there was another battle that occurred between the Vrishnis and Andhakas, and this second battle also took place in the same region. But one of the most fascinating things in these ancient Hindu accounts, is the description of the "flying craft" that were used in these battles. These craft were called "vimanas" and were utilised to launch weapons with fearsome destructive powers.

It seems that these vimanas, as well as being capable of flying around in the air, could also land on water. Part of the text mentions a hero called Adwattan who, after alighting on the water (quite a feat in itself), is said to have unleashed a terrible weapon of mass destruction that they named the "Agneya" weapon'. It was said that even the gods could not resist it, or had no defence against it. This weapon was described as a blazing missile of "smokeless fire" that when released, sent clouds roaring upward. This already sounds like the description of a nuclear bomb. But the text goes even further to describe the fallout of dirt and gravel that reigned down from above with fierce winds and an incandescent cloud billowing upward followed by darkness. All of this sounds very familiar to anyone who has witnessed a nuclear test detonation – just as is the thick gloom that was said to have descended upon the Pandava hosts after the explosion. All of this could not sound more like nuclear combat, each side toe to toe with their stated enemy.

The destruction in this ancient war is further described; the Earth was said to have been scorched by the violent heat of the weapon, elephants burst into flames and the waters boiled. The Earth shook and animals fell to the ground and birds croaked wildly. A person named as Gurhka, flying in his vimana, attacked three cities using his projectile that had all the power of the Universe. Whoever wrote these words must have been close enough to witness all this horror and destruction and if so, would possibly have died later from radiation poisoning. An incandescent column of fire with the brilliance of ten thousand suns is formidable indeed. It states that the corpses were so badly burned and charred that they were almost unrecognisable.

Everything that is described in these accounts brings to mind images of the aftermath of the Hiroshima bomb – images that could not be shown to the general public because they were so horrific. But the service personnel, who might at some point find themselves embroiled in such a scenario, had no choice but to view them. They were also obliged to take part fairly near to the test points to experience the effects of nuclear warfare as closely as possible in order to be as best prepared as possible in case it should ever happen.

It seems impossible that these scenes could be created solely from the vivid imagination of a creative writer – the comparisons with modern-day warfare are too numerous and accurate. Other factors that emphasise the nuclear aspect include stories of the victims' hair and nails falling out, birds turning white, foodstuffs being poisoned without seeming cause, and broken pottery. After the horrors of Hiroshima and Nagasaki, there were many lingering deaths and long-term, chronic health issues such as residual radiation and leukaemia. If anything good can be said to have come from these horrific incidents, it would hopefully be in the form of a lesson learned – a lesson that would ensure that such weapons would never be used in anger again. Yet unfortunately, as usual, humanity has learned nothing from them and now, even relatively unstable countries are in possession of these weapons.

Apart from the possibility that the references to what seem like nuclear war in ancient texts might have been the result of eye-witness accounts, another possibility that we might entertain is that these are the writings of a gifted clairvoyant vividly seeing a later nuclear age. Perhaps this might have been a clairvoyant or prophet similar to Nostradamus who seemed to have the power to foresee certain future events and even get as specific as coming close to naming perpetrators – such as his mentioning of "Hister" in place of Hitler.

Whether these texts contain accounts of past or future events, the similarity to modern-day warfare is surely something worth exploring, yet, despite the incredible similarities, intellectuals such as V R Dikshitar insisted that they were simply the result of an overripe imagination on the part of the ancient writers. I would argue in turn, that from a modern-day point of view, it could not have been anything other than a nuclear exchange. And, in our own age, physicists have recently begun to explore this rather astounding possibility and to acknowledge that something quite profound must have occurred to make the ancient historians record it all.

In our own age, it was round the year 1909 when physicists began to consider the possibility of harnessing the formidable power of nuclear energy for positive advantage. It is this that eventually led to the nuclear power stations that we have in existence today. However, even at this time, at the very beginning, some warned that with regard to radiation, a monster that could not be killed could be created; a monster that even today we cannot eliminate so we bury it in the hope that our descendants might be able to solve the problem.

It has been suggested, as mentioned in another work, that the offending material, in the form of nuclear waste, could be jacketed in titanium and launched into the Sun. It seems a logical idea to use the space shuttle for such ventures, including recovering some of the space junk that threatens the space station and of course the astronauts. However, the material would have to be encased in material with a much higher melting point than titanium, which would melt before reaching the Sun's surface. These are problems that persist. But in any case, it seems amazing to remember that the twentieth-century inventors of this technology were aware of these ancient Hindu texts – Robert Oppenheimer (known as the "father of the atomic bomb") was certainly one of them. He foresaw the positive as well as destructive use of nuclear energy.

The physicist Frederick Soddy, in his 1909 work *The Interpretation of Radium*, was certainly aware of the Hindu legends and made references to them, such as "Can we not read into them, some justification for the belief that some ancient race of men attained not only the knowledge that we have so recently won, but also attained the power that is not yet ours".

The author Rene Noorbergen said, "I believe that there have been civilisations in the past that were familiar with atomic energy and that by misusing it they were totally destroyed and that since 1945 of course, we

How Old is Technology?

have learned what the effects of the dangerous and destructive power of the atomic bomb are, and the descriptions that we have become aware of from ancient texts have suddenly become very real".

We have mentioned Robert Oppenheimer, who was one of the scientists involved in the initial tests of the atom bomb during the Manhattan Project. He was aware of the ancient writings in the *Mahabharata* and quoted from it thus, "I am become death the destroyer of worlds." The fact is that the group of scientists that envisaged, built, and tested the atom bomb could, in fact, have been the destroyers of this world; it was later admitted that they did not know for sure whether the atmosphere might ignite in some furious chain reaction and burn itself away.

They probably felt reasonably sure that it would not happen because of the gasses that make up the atmosphere, that is 78% nitrogen (an inert gas), 21% oxygen (inflammable), and 1% other gasses. With hindsight, we could question just how justifiable it was to develop such a weapon when most of the German cities had already been flattened and their top scientists were mostly fleeing to the allies. But some German scientists were also fleeing to Russia and herein may lie the clue as to why such a development might have been considered justifiable; Stalin knew of the bomb and although an ally at the time, Russia was seen as the enemy of the future.

Both India and Pakistan have the bomb now, and we can only hope that the legends that emanated from these countries do not repeat themselves as history so often does. Japan however, has decided to opt out of the development of nuclear weapons. It is perhaps not surprising that as the country that experienced the two most horrific nuclear attacks in history – Hiroshima and Nagasaki – they have wisely decided that they have nothing to gain by entering the global nuclear arms race.

The apparent use of atomic weapons in India 4,500 years ago could not have come about without a complete grasp of nuclear physics that would rival our own today. And there is indeed evidence of such

15

knowledge preserved among the ancient Hindu texts and records. For example, several Sanskrit books contain references to divisions of time – they cover a very wide range from immense lengths of time, down to the very minute. So why is this knowledge so important? Because these miniscule time periods could be related to the radioactive half-life of certain materials. Modern Sanskrit scholars have no idea why such large and miniscule time divisions were necessary in antiquity. All they know is that they feel obliged to preserve the fact that they were used in the past and are required to preserve the tradition to keep the legends intact, no matter what was written.

When we mentioned Tubal-Cain and the other Patriarchs with their advanced metal technology in 3,860 B.C., and then find that a couple of hundred furnaces were operating around 2,500 B.C. (as will be explored in a later chapter) – this seems proof of the legacy of Tubal-Cain. Even more staggering, if nuclear weapons were not only in existence but also in use around the year 2,500 B.C., then this makes the work of the patriarchs seem like a stage of advancement that was leading up to that point in history. It all points to something like a pre-arranged plan beginning with the information given in Genesis, particularly regarding all the other skills and crafts. If we estimate, from Biblical chronology, that Adam's birth would have been approximately between 4,800 B.C. to 5,000 B.C., then we ask again, how could flying craft and nuclear warfare have occurred in such a comparatively short time after the birth of the first created human?

We have suggested that the minute measures of time mentioned in the ancient Hindu texts could easily be related to a nuclear programme, which would require a working knowledge of the radioactive half-life and decay of certain elements during a nuclear programme. This knowledge should have taken very long time periods to have evolved. Consider the time it took from the Bronze Age to the 1944 Manhattan Project. This would have put the beginnings of such knowledge almost back as far as the very birth of Adam, almost as though technology was a gift as soon

How Old is Technology?

as man appeared. The only phenomenon in nature that can be measured in extremes of time – from billions of years down to the millionths of a second – is the disintegration of radioactive isotopes. The Hindu texts when venturing into cosmology, mention the "day" of Brahma, which is a period of 4 to 5 billion years. This is noticeably close, we can see, to the currently estimated age of the Earth, which is said to be 4 ½ billion years old.

And this brings us back to the Bible. When Moses was compiling Genesis, he refers to "days" when writing about the creation of the world and all the creatures in it. This provides a possible answer to the puzzle; instead of his "days" consisting of twenty-four hours, he was possibly referring to a different measure of time when he used the word "day". Nevertheless, people in his time must have known what a "day" meant to them, particularly his Egyptian tutors. To return to the Hindu pre-occupation of dealing with miniscule time periods, they could only be related, as said, to radioactive decay. Uranium 238 has a radioactive half-life of 4 ½ billion years, but it is quite staggering, to say the least, to think that the ancient Hindus could have been familiar with all the subatomic particles, such K-mesons and hyperons with their radioactive half-lives of millionths or even billionths of a second.

However, the written Hindu chronicles and their mind-stretching contents depicting what seem like nuclear weapons being dropped from flying vimanas, are not the only clue that leads to the possibility of ancient nuclear combat. Physical evidence of nuclear combat seems also to have been found in the areas, such as the Upper Ganges river in Northern India – the same area that was mentioned in these texts. Here have been found a considerable number of charred and severely burnt ruins. After close examination, it seems abundantly clear that the ruins were not burned by ordinary fire. Ordinary fire would have a distinct signature. Instead, here can be found masses of stonework that have been fused together creating deeply pitted surfaces that look quite like molten steel that at some point bubbled before cooling.

Further to the south of this site there are more of these ruins; here can be found walls that have been glazed and split apart – becoming vitrified, melted and then crystallised – obviously by a tremendous heat source. It has been determined that only the enormous heat released by an atomic blast could have done this. Moreover, a Russian researcher, A. Gorbovsky, noted in his Riddles of the Ancient Past that among these ruins a human skeleton was discovered and after certain tests was found to have fifty times above the normally expected background radiation count.

It is not only in India where these signs of immense heat, possibly due to past nuclear conflagration, are to be found. Researcher Erich A. Von Fange saw indications of a past nuclear holocaust in a place not far from ancient Babylon. Here, he discovered an area that contained similar signs as found in India. He found the melted ruins of a ziggurat with brown and black masses of brickwork that had been transformed into a vitrified state and been completely molten.

Another place where an unmistakable nuclear signature was found was in Israel when archaeologists were excavating in 1952 and found a layer of fused green glass that was spread over an area of several hundred square feet. This fused quartz sand was very similar in appearance to the layers that were found to have changed to such greenish glassy material in Nevada USA. This effect in Nevada was caused by the tremendous heat of the nuclear test explosions carried out there. They called it "trinitite" (after the trinity test site). Other signs or evidence of a nuclear event have been found in Southern Iraq (again near Babylon), and it does not stop there. The same signature of a widely spread area of nuclear conflict is also to be found in the Southern Sahara region and also in the remote and most desolate areas of the Gobi Desert in Mongolia. Evidence of enormous heat have also been found in China, quite near to the actual atomic test site that was used later.

When scientific tests were carried out upon these sites, the composition of the fused green glass from the different sites, matched each other. However, it must be commented upon that while scribes in India seem to have been busily recording what seem like historical war records for their successors, there is a lack of such detailed descriptions from other countries with possible ancient nuclear histories. So are we to assume that only India had these weapons as a single nuclear power? Or perhaps we have simply not yet found this evidence.

In *Pillars of Fire*, I speculated on the destruction of the "cities of the plain," that is Sodom and Gomorrah, and wrote that this destruction was possibly a nuclear obliteration carried out by a higher power. In the Bible, it was clearly observed from a high point, by the patriarch Abraham, who was standing alongside one of the perpetrators who was accepted by Abraham as an "angel". He saw the funnel of smoke arising that could only be described as "like that of a furnace!" Lot and his wife and daughters left it a little late when fleeing to the hills to avoid ground zero, but Lot's wife paid with her life for her disobedience and curiosity, and looking back, was turned into a "pillar of salt."

In the light of all these discoveries, it is hard to accept that anything else but the tremendous heat from a nuclear blast could be responsible – if only because nothing else could produce the heat required to melt all that rock. If it was just one area with the remains of the eroded crater with fused material, we could offer an alternative explanation, such as asteroid impact.

It is hard to accept that such a widely spread nuclear exchange could have taken place so long ago, but we cannot avoid the evidence, and have not yet come up with an acceptable alternative explanation. I suspect that few historians would accept it, but rather accept the "colourful imaginative legends" option. Nevertheless, the atomic signature is there. For it to have been a worldwide event would seem even more incredible, but we cannot simply ignore the evidence; there are even vitrified walls that have obviously been affected by a similar extreme heat source, evident in the pre-historic forts in Europe and in the British Isles.

THE MODERN ANCIENTS

In the Lofoten Islands, off Norway, there are more vitrified walls to be seen, Scotland also has them. They are also evident along the coast of Ireland where granite fortifications have been subjected to such heat that they have been melted to a depth of one foot. The mountains and lands of Tibet are not extremely far from India and certainly, the ancient memory of the Tibetan *Stanzas of Dzyan*, whose origins date back several millennia, appear to verify the Hindu legends of an apparent nuclear holocaust that engulfed the warring nations. This text also mentions that they used flying vehicles that had the capability to release fearsome weapons.

And, there are yet more areas of the world affected. The continents of the New World also pose quite a few examples of ancient pre-historic cultures that show all the signs of having been destroyed by nothing less than the enormous heat produced by nuclear weapons; we know of no other conventional weapons or armaments possessed in ancient times that were capable of leaving behind such a trace. Not far from Cuzco in Peru, fairly near to the pre-Inca fortress of Sacsayhuamán, an area of some eighteen thousand square yards of mountain rock has been fused and crystallised; this is not only evident on the mountainside, but it has also affected a number of the dressed granite blocks that were used in the building of the fortress itself. They also show obvious signs of quite similar vitrification, clearly caused by immense highly-radiated heat.

And so we ask again, why with so much seeming evidence of fearsome worldwide warfare, was it only meticulously recorded in the annals of the ancient Hindu texts? Perhaps we should ask ourselves how well such records could be kept; a vast number of records could be destroyed in one fell swoop, as they were, for example, in the great libraries of Alexandria where they were quite vulnerable to the destructive rampages of various conquerors. Of course, the South American historical legends and history were also subjected to such treatment through religious fervour by the bishops and priests that accompanied the conquistadores – all of whom combined, over the years, to destroy physical as well as oral culture.

We will travel a little further north in the Americas and present even more evidence of this extraordinary heat signature that has been found on ancient ruins. In 1850, the explorer Captain Ives William Walker saw a number of ruins in the Death Valley, California. He discovered a city stretching for about a mile, and he noticed that the outlines of the streets and also the locations and the positions of the associated buildings, were still quite visible to him. As he moved around the centre, he found a large structure, the southern side of which clearly displayed the same evidence of melting and vitrification that had been found in the other locations mentioned here.

There are also many other areas in America – in parts of Southern California, Arizona, Colorado and in the Mohave Desert – where patches of the same fused green glass have been found. So what was the cause, or reason for all this worldwide horror? What had all these countries and locations done to deserve it? What kind of retaliatory forces or weapons did they possess? In their legends (as far as we know to date) there is no mention of anything like the vimanas fighting back with flaming darts of destruction and no heroic characters seem to have emerged from the horror. It seems to have been world conquest for world conquest's sake, or perhaps for the other countries' valuable resources.

It is self-evident that with all these ancient signs of nuclear aggression, there must have been a great mass of people all around the world that would have been seriously irradiated. Although skeletons have been found with evidence of having been subjected to extraordinary levels of radiation, they should in theory have been found whenever diggings took place – either for foundations or during any purposeful archaeological excavation. Many bones have been found, but has anyone ever thought of calling in experts in the field of nuclear physics at the same time that routine notification of the police occurs? How many highly radiated bones have been found and later routinely disposed of without any realisation of this possibility? Or, did this destruction occur so long ago that these bones would now be fossilised?

If this terrible conflagration did in fact occur on a worldwide basis, then alongside the quick deaths that would have occurred, many others would have experienced painful and long-lingering symptoms for which death would have been the only blessed release.

I wrote in *Pillars of Fire* that the wiping out of the two cities of Sodom and Gomorrah had all the earmarks of a nuclear obliteration along with the usual evidence of the fused green glass that was found there. But the point is that if radiation was spread out from ground zero, some people further away from the event may also been affected. I contemplated in the above-mentioned book on the question of whether some of the many "lepers" that seemed to exist in ancient times (to make up entire colonies) may have been in fact affected by these events, as were perhaps also those people that were described as being "stricken with the palsy." Today, leprosy is a disease that we feel we have conquered, but it seems to have been fairly widespread in Biblical times.

In the story of Sodom and Gomorrah, the mysterious "angels" that carried out the destructive operation to annihilate the cities of the plain must have been aware of the effect of nuclear fallout as they advised Lot and his family to "head for the hills," or a high point and find shelter in a cave. Perhaps all the events we have mentioned, contributed greatly to the extremely noticeable drop of human longevity evident in the post diluvian peoples. The dates and ages of the patriarchs would have meant that some of them and their descendants would have been living when the events were taking place.

Clearly, as said, a sound knowledge of nuclear physics and atomic theory would have been necessary in order to develop the awful weapons written about in the Hindu texts. Yet, all this knowledge seems to have disappeared along, perhaps, with the people most qualified and able to produce the horror of their discoveries; it all seems to have disappeared into history along with the warlike activity itself. But history repeats itself, and ancient knowledge seems always to revive and to be studied all over again up to the point that we are experiencing today with so many countries possessing nuclear capability.

In our own era, we assume that Democritus (460–361 B.C.) was the first to formulate atomic theory (or perhaps we should say, rediscover), but being a scholar and proficient in the knowledge of other subjects, Democritus would surely have been aware of many aspects of ancient history. He would have been aware of events that occurred thousands of years before him – even the alleged Hindu wars and possible nuclear exchanges that we have mentioned, would all have been knowledge that was available to Democritus. However, whereas Democritus and his actual existence is factual, the Hindu legends have not proven to be so because the only reference that we have to them is in the written accounts in the aforementioned *Mahabharata* and other works.

The ancient Greeks wrote of many things and of course, their wisdom was obtained from the ancient Egyptians who preceded them and were said to possess a very ancient historical record of their own long history. However, the Greek writings do not seem to mention any factors relating to the conflagrations evident in all the buildings and edifices that we have mentioned. Perhaps the Hindu people were so protective of their ancient historical writings that they did not allow them to be released to the world. Or, perhaps it was just a matter of disbelief such as is evident in the comments on the Hindu works that have been made by scholars such as V. R. Dikshitar who spoke of these works in an almost embarrassed tone – stating that they were all only imaginative tales. However protective the Hindus may have been with regard to their written legends, they could do little about the observable evidence in the form of vitrified stonework that has been found all over the world.

To mention Democritus again, he seemed to anticipate the view of modern physicists when he said (2,500 years ago) that "in reality, there is nothing but atoms and space." Our previous assumption that Democritus would surely have been aware of all those startling Hindu legends may be proved correct as we learn that Mochus the Phoenician was said to have communicated this information to Democritus. Moreover, the conception that Mochus held, with regard to the structure of the atom, was in fact nearer to the truth than what Democritus believed – in that it was not the smallest structure simply because of its divisibility.

The ancient Greek philosopher mostly claimed that in fact, there was no distinction in kind between the stellar bodies and the Earth. Because he stated that everything was atoms and space, we can assume from this idea that by "stellar bodies" he meant the stars and the planets and all the other rubble floating about in space. Of course, we now know that this is in fact so and that even our own bodies are made up of this same substance. It would be quite interesting to know if Democritus realised this as the sciences of biology and genetics had surely not progressed that far in his day.

Yet some may ask "had it?" After all, before its destruction, it is said that the great Alexandrian library had operating theatres and biological dissection rooms and laboratories including facilities for the study of botany for example and everything learned was written down and logged in scrolls that numbered in their thousands – all later to join the flames of destruction.

Other Greek scholars spoke along similar lines as Democritus – scholars such as Leucippus and Epicurus who no doubt contributed their own knowledge in more re-enlightening scrolls that were logged away with all the other works. Like the Greeks, the Romans, in their turn, learned much from those who went before them. For example, the Roman scholar Lucretius (of the first century B.C.) wrote about atoms rushing everlastingly through space and sometimes, as he put it, colliding and undergoing myriad changes due to the disturbing impact of collisions.

The accuracy of this thinking seems incredibly advance when we consider the many centuries that had to pass before our physicists actually observed these collisions in special liquid "bubble chambers". Further evidence of the advanced knowledge that existed in the time of Lucretius can be seen in his statement that "it is impossible to see the atoms because they are so small."

Lucretius and many other learned Romans must have been aware that their military leaders, during their continual conquests, were responsible in some cases for the destruction of much written data. Nevertheless, the learned Romans must be commended for preserving as much of this knowledge as they could, and for not fearing to quote the theories that they had likely inherited from earlier sources. And they must also be commended in the event that some of their ideas and theories emanated from their own minds – and for not fearing derision or scorn from those unable to grasp their advanced ideas.

But we return to the idea of original tutors, and to the question of who tutored the tutors, where did it all begin? What led the great Greek and Roman scholars to even envisage tiny invisible particles like atoms? One can imagine scholars being aware of air as a substance – as something tangible that they could feel as they wafted a hand quickly through it. So if they were aware of its existence, then of course they would be curious to learn about its composition. Yet it would have been a massive leap forward to arrive at the conception of atoms, especially as (we assume) they had not yet learned of the air's composition and of the gasses of which it is comprised, namely nitrogen, oxygen etc. It is rather like the conception of the "chicken and the egg" scenario when we start to consider who the original learned ones were; and when we ask this question, we move into the realms of New Age philosophy with regard to the possibility of aliens visiting the Earth during some remote period in our past and busily teaching humans their vast array of knowledge.

Yet, the question must be asked. This knowledge had to start somewhere and at some point in the distant past, but it is only humans or perhaps "humanoids" that were, and are, concerned with all of this. Certainly not the apes, fishes, insects or other animals. It is only of concern to those who possess the ability to wonder or be concerned about the world around them. That is, ourselves, plus any other advanced and intelligent species that surely must exist somewhere else in the Universe.

Are we now approaching the question of the possibility of some grand creator? We do not intend to pursue the question of who created the creator, but ordinarily, knowledge is gained by those who stand on the shoulders of people who have gone before them. Knowledge does not simply enter one's head, but hunches, ideas, and the consideration of possibilities, certainly do and these are built upon via research until a tangible idea or theory emerges. This is of course, the whole basis of science.

With ever-increasing frequency, new hypotheses are tested, current scientific theories modified, and new formulae constructed – mostly with the aim of confirming or proving previously accepted theories. However, sometimes bits of disconnected information continue to pierce the walls of scientific complacency and as a result are often ignored. This is certainly true, with regard to those irritating "ooparts", that is those "out-of-place" (in this instance technically constructed) artifacts. The scientific community is in general reluctant to go anywhere near them.

To mention Lucretius again, in his *On the Nature of the Universe*, he states, "There can be no 'centre' of infinity." However, one could possibly argue against this conception, for example, if we accept the theory of the so-called "Big Bang" that scientists (and theologians) accept as the starting point of everything. Science and religion could at least agree on the point that in the beginning, creation of a scientific or divine kind occurred. The way that it is instilled in our minds is that from that point onward, matter spread outward toward infinity – this would be rather like a hand grenade that explodes say six feet off the ground – any soldier standing in any point around 360° would be at risk of injury or death by shrapnel. So, the spreading of the Universe would have originated at the point of the explosion – and this would be the "centre of infinity" that Lucretius could not envisage.

How Old is Technology?

Strangely, the explosion shows no sign of slowing down; in fact, quite the reverse. Although it reflects quite badly on human behaviour patterns, every new invention is initially assessed for its potential value in warfare. The ancient sages knew of this and wrote of the dangers of revealing knowledge "to those who might use it for destructive aims." A thousand years ago, a Chinese alchemist who experimented on chemical discoveries stated, "It would be the greatest of sins to disclose the mysteries of your art to soldiers".

Nevertheless, someone must have leaked the information to soldiers because it is well known that the Chinese used chemical explosives such as gunpowder and rockets. But some inventions can of course be developed for use either in peace or war – for example, if it had not been for the discoveries of explosives and chemical reactions, men would never have stood on the Moon.

When we think of the Hindu texts and the potential harnessing of nuclear power thousands of years ago, it should not really surprise us when we read in the Brahmin texts statements such as, "There are vast worlds within the hollows of each atom, multifarious as the specs within a sunbeam." How could they possibly be aware of the existence of subatomic particles and not possess the advanced and technical apparatus that our physicists of today possess? It sounds as though they envisaged the atom being similar, in a sense, to a star surrounded by its whirling planets (which it is).

Not all of the incredible warlike and destructive events that we are invited to believe took place in India, and that involved the Hindu Gods hurling what seem to be nuclear weapons at each other, would have been I.C.B.Ms, or intercontinental ballistic missiles, but some must have been. This is evident from all of those melted stone structures found elsewhere in the world – and how were they delivered? Enter the vimanas.

Just like modern aircraft, the vimanas were designed for both military purposes and for the ancient peoples' version of civil aviation. It is hard to imagine that the practice of aviation was real and actually taking place at the very dawn of history, but we get indications that this was so from terms in ancient Hindu texts such as "vimanas vidya", which refers to the science of building and piloting flying aircraft.

How can we accept that at such an ancient time period, especially if we think in Biblical terms as per Genesis, the appearance of Adam, and the early antediluvian patriarchs, that such technology existed? The classic Indian epic knowns as the *Mahabharata* has already been mentioned here and it is said to be one of the oldest books in the world. Within the text of this book there is a clear reference to what was called an "ariel chariot". It was described as having sides of "iron" and was clad with wings. Most certainly the wings make sense but in terms of aviation, where weight is of paramount importance, clad with sides of iron makes no sense at all. However, it is entirely possible that this was just a misinterpretation and that all that was meant was that the sides were of metal composition.

The Hindu epic, the *Ramayana*, describes the vimanas as a double-floored craft, circular in nature with fitted windows and an upper dome. We have to mention at this point how similar this description is to many modern-day reports of the unidentified craft we describe as UFOs. This craft may seem large and heavy in our imaginations but it was said to fly with the speed of the wind and gave a "melodious sound" (it is quite certain that many people who now live below flight paths into airports would much prefer a "melodious sound"). The text mentions in particular that not just anyone could be entrusted with piloting these craft; it is said that a pilot had to be trained well or he had little chance of being given the privilege of flying one.

Another comparison that could be made with regard to modern flying craft is that these vimana could carry out manoeuvres similar to a helicopter in that they could stop or hover and remain quite motionless in the sky. Yet, as they had no accoutrements such as large blades to accomplish this we must assume that these craft had the ability to perform like the Hawker Harrier Jet. In the ancient book mentioned, there is an

How Old is Technology?

account of what seems rather like the pilot's or passengers' eye view from the flying craft; it states how the Vimanas soared above the clouds and from on high the ocean was described as "looking like a small pool of water," clearly, the craft was at a very high altitude. This text goes on to relate that the aviator (and the passengers of course) were able to see the coast of the ocean and the delta of the rivers. This sounds very much like the craft was flying over a major delta such as the Nile Delta in Egypt.

Just as today, at an airport when aircraft are not flying but having maintenance carried out, they move into hangers; and this ancient book that gives all this detail did not omit to mention that the craft were kept in hangers called "vimanas griha". They even went on to describe the fuel that was used to power these craft – it was described as a yellowish white liquid.

All the above detail was also mentioned in a book by C. N. Mehta, which he titled *The Flight of Hanuman to Lanka* published in Bombay in 1940. Again, just as today, these aircraft (or vimanas) were constructed for and were used in warfare as well as travel and leisure. This last use implies that a person of adequate means could, if he or she so wished, own their own personal vimana, much like a person or business mogul of today might own a private jet. The quite precise and exact detail that was found in the ancient texts seems to reinforce the idea that these texts present valid and factual data rather than fantasy.

On the other hand, another Indian work called the *Panchatantra*, at least in parts, tends to sound more like a novel or an imaginative story. Yet this text also has parts that may be factual. This ancient Indian story relates that six young men constructed a flying craft that could take off, fly and land. It had a built-in and rather complex control system that provided a safe, fast flight with excellent manoeuvrability. Today we can envisage a group of enthusiasts who are proficient in model aircraft design, construction and radio control expertise, constructing large flying craft such U.A.C.'s (unmanned ariel craft, i.e. drones). Some enthusiasts do indeed do this, and one might imagine that the next step might be to construct a flying machine that one could enter, cause to take off and fly. Such a machine would surely have to be produced, tested and flown with the aid of an aircraft factory. So this could be read as a factual record of an advanced but long-lost technology.

Another ancient text, the *Samarangana Sutradhara*, provides a seemingly factual account of air travel from every angle. This poetic treatise is said to contain some 230 stanzas that deal with the construction of flying machines. It speaks of take-offs and landings, normal and "forced" landings, cruising flights that cover thousands of kilometres, and even speaks of the common hazard that both civil and military aviation faces which is bird strikes. The latter would of course not be a problem if the craft, rather like a glider, was just floating through the sky. So, this mention of bird strikes indicates the speed and rapid velocity of the craft being described without which the birds would not present a problem.

These stanzas also mention possible collisions with other flying craft, which again, would not be much of a problem without high-speed velocity. These works display quite clearly that all the issues, problems and dangers that affect modern aviation today, were also present in those ancient times.

All this seemingly factual data is quite difficult to disregard as fable, or the product of wild imagination on behalf of the writers – especially in the twentieth and twenty-first centuries when the aviation written about in the ancient texts so closely resembles our own. Moreover, as well as the military air battles and raids that took place, it appears that these ancient people even progressed to manufacturing and delivering biological weapons and possibly chemical weapons. The "Samara" was a missile that was said to "cripple" the enemy. And another weapon named the "Moha" apparently induced in its victims a complete state of paralysis.

Some scholars believe that the Hindu sacred texts, including the *Samarangana Sutradhara*, were compiled in the eleventh century. However, others argue that they were only "updated" or "amended" at this time and maintain that the source of these works dates far back into unknown antiquity; this latter argument gains traction if the texts can be related to passages in other works that have been proven to be ancient – which they do. Some of these texts even include what seems like aviation "trivia" such as what clothing the pilots wore – making it seem even more unlikely that this was merely "drama" on a grand scale.

There is an Academy of Research into all these ancient events in Mysore, India, where an in-depth study and translation was undertaken, the final results of which were compiled into a work titled *Aeronautics, A Manuscript from the Pre-Historic Past*. After the publication of this document, and the immense amount of information that it contains, no one can possibly still support the supposition that these ancient accounts are nothing but imagination. The view of the interpreters and translators of this document, is that there are within it so many revelations of compelling and detailed aircraft design, performance flight, and function that they completely rule out any consideration that they are simply the product of someone's imagination.

Within this document are many details such as "the aircraft can go by its own force [an engine] and fly like a bird, operate on the water or in the sky. This craft is called by the Sages of Old a "vimana" or that which can travel in the sky from place to place." This text makes it clear that the body of the craft must be strong and durable and built of light wood. However, we must pause here in order to note that there appears to be a contradiction here with regard to the description given in the *Mahabharata* of the flying machine that was described as an ariel chariot with sides of "iron" (or as we previously suggested, "metal" of some kind such as aluminium or magnesium). In WWII, some aircraft were clad in both wood and metal – wood being used widely, for example, in the DeHavilland Mosquito. Some aircrafts in the modern day have also utilized fabric. The light wood mentioned in the compilation from the Mysore Academy was named as "Lachu-Daru".

Although wings are mentioned as a necessary feature and the craft is said to be like a bird in flight with its wings outstretched (mahavinhanga), circular craft were also described. The fact that "iron" is mentioned in this text, along with other heavy apparatus, is an idea that anyone with an extensive background in aviation would find baffling. The text mentions that "in the larger craft (daru-vimana), because it is built heavier, (alaghu), four strong containers of mercury must be built into the interior. When these are heated by controlled fire from the iron containers, the vimana possesses thunder power through the mercury." So when reading these texts, anyone proficient in the knowledge and idiosyncrasies of aviation, and the pre-requisites that are necessary to allow it to happen, must have their interest aroused in one moment and then deflated in the next when confronted with such obvious contradictions.

Doubt increases when one focuses on the weight factor as the story continues: "The iron engine must have properly welded joints to be filled with mercury, and when the fire is conducted to the upper parts, it develops power with the roar of a lion. By means of the energy latent in mercury, the driving whirlwind is set in motion, and the traveller sitting inside the Vimana may travel in the air, to such a distance as to look like a pearl in the sky." With all the mention of "fire" in this process, one might not wish to get too close to one of these craft when starting up – they must have had very stringent safety precautions in place! But with all the heavy-sounding apparatus mentioned, the craft would certainly have needed all the "thunder power" it could get to even get off the ground.

Because this historic data goes into such detail on the one hand, but then introduces doubt and uncertainty with regard to its validity on the other, we must consider whether all the words in these ancient texts have been interpreted or translated correctly or not. For example, when interpreting Biblical texts, the Hebrew language can be very difficult to translate correctly as one word could mean many things. Examples in English include the word "bow" to mean paying homage, "bow" as it relates to archery, "bow" as a violin bow, or "bough" as in the limb of a tree.

Rene Noorbergen states: "In these archaic accounts, conspicuously missing from these ancient texts, is any distinct description of how vimanas were actually constructed." This is true in the sense that although the actual materials used in the process of construction of the craft are mentioned, the technical building method from initial assembly to the completed product is not described in detail. One explanation for this lack of detail might be that there was a reticence to include too much of what we might regard as classified information. In fact, this explanation is made apparent by the writers of the sagas themselves, who wrote that any person not initiated in the art of building these machines of flight might cause "mischief" – so clearly it made sense that this information was classified. It also shows that industrial espionage was already in existence during these times; and if the vimana were used in warfare, then the possibility of espionage and the need for secrecy makes sense.

Yet, as usual, a contradiction arises. As already mentioned, the vimana were used for civil air travel as well as for warfare and moreover, any "person of means" could own a vimana for private use. This would have meant that anyone with advanced knowledge could (to use a phrase in modern parlance) "back engineer" a vimana and soon discover the construction details and engineering required to build one – just as is maintained today by ufologists with regard to alleged UFO crashes.

Nevertheless, the implications are that in the main, a sort of control of this knowledge was maintained by a select few. And any industrial spy who obtained these writings would surely have had some trouble when trying to sell them to the highest bidder when that bidder heard the idea of flying machines being "clad in iron" with "large pots of mercury" and fire being needed; the spy would surely have been derided and mocked as the recipients would surely have asked him how he expected such a weighty contraption to "soar through the air." But maybe this was the point – perhaps these details were purposely written to be misleading, to hide the real facts, especially when we recall that "anyone not initiated in the art of building these machines will cause 'mischief'." The key point here being the idea of "initiation".

I will now reveal an account revealed to me by the Chief Inspector in Engineering and Quality in an aircraft construction company when I accepted a post in his department. During the early days of Concorde construction at Bristol, there were two inspectors that joined the company and who were under surveillance, in part because of their background and immigration status.

The function of an aircraft inspector depends on the department he is assigned to; it may be the sub-assembly area where, just as in ship construction, a sort of keel is laid and the frames of the fuselage are fitted with so-called "stringers" that make up the length of the fuselage. This fuselage is eventually clad in sheet metal (not iron) and all the necessary parts of the structure fitted into it. Eventually, the wings and the tail structure are added. An inspector has to examine every part of the structure and pass it as fit for assembly. Then working with specific blueprints, he uses his official rubber stamp, with his number embossed on it, to approve the record card regarding that particular "operation". These "tasks" continue as the aircraft moves slowly along the production line or "track" just as in a car factory, and eventually these so-called "tasks" add up to the complete build.

The work can be very repetitive and the drawing office staff soon get well acquainted and friendly with the inspectors coming in to draw blueprints for their build tasks. Nevertheless, when an inspector attempts to draw blueprints for operations outside his allotted routine tasks, suspicions are aroused. Sometimes, the drawings requested by the inspectors that were under surveillance, were not returned at the end of the day. The operation then apparently began to ensure that only purposely-produced drawings were issued to these suspects – drawings that gave false information on factors such as the "gauge" or thickness of the metal, the type of metal, and the spacing of the rivets etc. If all of these false specifications were ever employed in construction, then the aircraft would not stand up to the stresses and speeds of an aircraft such as the Concorde.

So all the data that these suspects obtained was outside the safety limits of construction – and the then "Eastern Bloc's" copy of the Concorde, nicknamed "Concordski", was therefore doomed from the start. The inspector who related all this information to me has unfortunately since passed away. He may have been embellishing these accounts; but it was said that some of the people who were involved in monitoring this case of blatant espionage were present when the "Concordski" disintegrated at an air show, knowing what was likely to happen. Needless to say, it never went into production.

Unfortunately, not all cases of industrial espionage are discovered, but to discover that these cases may have been in operation thousands of years ago is startling. If we accept that these vimanas did actually fly, perhaps the real facts regarding their power sources and building processes were so secretly guarded and preserved that they are still to be discovered. As the Russian poet Valery Bryusov wrote in one of his poems, "the poets and sages, guardians of the secret faith hid their lighted torches in deserts, catacombs and caves." Perhaps the discovery of the Dead Sea Scrolls in the remote caves near the Dead Sea, hints at something that is only the tip of the iceberg.

One of the strangest factors in these accounts is the frequent mention of "mercury" – something that we know of as a metallic element and which is puzzling to imagine being used as a veritable source of energy or power. Has science made any attempt to explain this oddity? Is there perhaps an energy source related to mercury that we don't know about?

One study that explored this frequent mention of mercury and its possible use as a power source, noted that propulsion via the latent energy present in mercury was the chief puzzle concerning the Hindu vimanas as they are described in the *Samarangana*. It is interesting to note that the element mercury had a special place not only in the science of the ancients but also in the science of the many alchemists that existed in Medieval Europe as well as all over the world. The British nuclear physicist, Edward Neville da Costa Andrade, in a speech delivered at Cambridge, noted that the famed discoverer of the laws of gravity, Sir Isaac Newton, seemed to know something about the secrets of mercury. Quoting Lord Atterbury (a contemporary of Newton) Andrada said "Modesty teaches us to speak of the ancients with respect, especially when we are not very familiar with their works." Newton, who knew these ancient accounts by heart, had the greatest respect for them and considered the ancients to be men of genius and of superior intelligence. These were people who carried their discoveries in every field much further than we suspect we have today – judging by the very little of what remains of their writings. More ancient writings have been lost than have been preserved and perhaps our new discoveries are of less value than those that have been lost.

Edward Andrade speaks about Newton especially because of Newton's comments criticizing those who were too generous in disclosing certain alchemical practices of the time – in particular certain practices regarding mercury: " … yet because the way by which mercury may be so impregnated, has been thought fit to be concealed by others that have known it, and therefore may possibly be an inlet to something more noble, not to be communicated without immense danger to the world."

Is it possible to imagine that there is something quite profound yet to be discovered by our own scientists with regard to mercury? One cannot help but wonder what it is about mercury that could be of such "immense danger to the world." We know of course, that it is poisonous, but it seems that there is more to the issue than that; it seems apparent that the ancients were ahead of us in their knowledge on the practical applications of mercury.

Another indication of the importance of mercury comes in the form of a strange report that emanated from Russia. It appears that a group of

How Old is Technology?

Soviet explorers, when excavating a cave near Tashkent in Uzbekistan, discovered a number of ceramic conical pots – each one of them had been carefully sealed, and each one contained just a single drop of mercury. A full description, together with the appropriate illustrations of these mysterious pots, was published in a soviet periodical in a report titled "The Modern Technologist." But no consensus of opinion was reached as to what these strange clay containers and their mercury drops were used for. Ancient clay pots that were found in Iraq, turned out to be ancient batteries that generated a couple of volts each when a suitable electrolyte was used; it was assumed that linked together in a series they were used for electroplating metals. But mercury pots containing just one drop of mercury present an entirely different problem.

The Uzbekistani pots must have been used in those ancient times for something beyond our present understanding. While a purpose could be imagined for the pots that were discovered in Iran, nobody has yet been able to suggest a similar use for the mercury pots. The ancient Indian texts talk about large amounts of mercury – so it staggers the imagination to think that if just one drop of mercury might be used as a power source, then how much power did the vimanas possess when being powered by a couple of cauldrons of the stuff? The accounts of fire that are mentioned are also a source of mystery in that it is difficult to imagine how they were controlled.

The mercury pots are but one of the many ancient artifacts discovered that seem to indicate a technology that is lost forever. When we look at items that crop up from Biblical sources such as the Old Testament, some interesting information comes to light; for example, we could ask, did the ancients have oil refineries in 900 B.C.? The biblical prophet Elijah had in his possession several jars of what could either be a volatile liquid that looked like and had the same viscosity as water, but could only be a bi-product of oil. It was either naphtha or petroleum. Elijah used it for a dramatic demonstration before the priests of Baal and the King of Israel, Ahab. We will come back to this incident in due course.

Petrol is indeed a bi-product of oil, but it is only one of many derivatives of that substance. To be produced, it has to go through a process of refinement, as do the other by-products. So where might Elijah have obtained petrol? Elijah was a dominant figure in the Old Testament and carried out many notable deeds. He could heal the sick and even brought a young child back to life. He had a special garment called a "mantle". This is a word still used today in the German language meaning garment or "coat". But Elijah's mantle had powerful abilities; it could part the waters of a river, a purpose that Elijah indeed used it for. Elijah's "mantle" was eventually passed on to his Disciple Elisha just before a craft descended. This craft was described as a "whirlwind" that took Elijah off to "Heaven" – a place beyond Earth.

I wrote about this event in *Pillars of Fire* – an event that occurred in full view of his friend Elisha. When considering the interpretation of biblical angels, I consider this as one of the earliest recorded abductions by creatures not of this Earth. Elijah, who appears in the Old Testament Book of Kings, has been likened to a being somewhere between John the Baptist and Jesus himself because of his special deeds and abilities. And he has therefore secured his place in biblical history.

Like John the Baptist, Elijah did not bedeck himself in "fine raiments" but instead dressed rather plainly (possibly in tailored animal skins) yet he was still considered a great prophet and healer. He used his aforementioned jars of petrol in an astounding demonstration in support of his belief in the one true God of Israel, Yahweh, over the gods of Baal who he denounced. The sceptics of biblical miracles refer to them as nothing more than clever tricks (and such magic tricks are as old as history itself). Perhaps in some cases this may have been true; the "miracle" performed by Elijah was certainly a "trick", but one with a good intent.

How Old is Technology?

During the reign of Ahab, there was much licentious and lewd behaviour among the population and the among the courtiers of the King. Elijah on the other hand was considered a chaste and righteous man. Because of this, Elijah made many enemies not least the King of Israel himself; King Ahab was irritated by Elijah and although fearing his alleged powers sought ways to kill him. The King and his couriers felt that Elijah cast guilt upon them via his righteous ways.

The King and a good portion of his subjects worshipped the God known as Baal. One of the reasons that King Ahab did not immediately have Elijah arrested and killed was that at this point in time a severe drought had fallen over the land and Ahab thought that he might use Elijah and his alleged powers to alleviate the situation as his own priests seemed to be having little success. Ahab no doubt wondered that if Elijah's God was as powerful as Elijah preached, then he might be able to solve the problem. It was at this point that Elijah decided to perform his trick. He did not lie and say that his performance was a miracle of God, rather, he let the people form their own conclusions.

Elijah likely decided that if his trick worked, and if people repented their wicked ways and turned back to Yahweh as a result, then the end simply justified the means. Elijah began the process by requesting that King Ahab's priests build a fire of staves but not to light the staves straight away. Elijah then built his own fire, but whereas the shallow trench dug around the priests' fire was filled with kindling, Elijah left his trench empty. At this point, Elijah ordered that an animal be butchered and the pieces placed on each of the fires. He then bid the priests of Baal to summon their Gods and have them start the fire. After sometime, the crowd began to shuffle and show signs of restlessness as they wondered where their Gods were. Why did the Gods not respond as asked?

It was then that Elijah opened his jars of liquid and poured them into his trench. At first the King and his courtiers began to laugh – was this Elijah going to use water to light a fire? But then their amusement turned to anger; how dare anyone pour all this water into a trench when such a severe drought had befallen the land and when every drop was so precious? In the heat of the sun, fumes were rising from the petrol and it was close to combustion. Elijah called upon his God Yahweh to display his power and show the people the way back to worship him as the one true God. With some kind of device, probably containing a flint, Elijah caused a great conflagration of flames, which astounded the people. After this display, many people turned away from Baal and Baal's priests – and many of these priests had their throats cut down at the riverside after the event.

Clearly, Elijah gained more power, respect and converts after this demonstration – and it is perhaps this latter point that later earned him his trip to "planet" Heaven. But before this happened, Elijah was to gain even more respect by also solving the drought problem. I wrote about this in *Pillars of Fire*; the people observed a craft that was far off over the water and described it as "no bigger than a man's hand." Later, this craft delivered a huge amount of water over the land and thereby solved the drought problem. It seems that some form of desalination process coupled with some kind of "cloud seeding" process was used (as it is today) to cause the rain. So while Elijah's petrol demonstration might be dismissed as "trick" – solving drought and other of Elijah's good deeds such as bringing a dead child back to life, can't be dismissed as such.

But the question remains as to where Elijah obtained his combustible liquid or petrol; it must have been produced somewhere in large quantities if Elijah was able to fill up large jars.

CHAPTER II

THE FACTORY ACT 2,500 B.C.

If we are to envisage a huge factory complex where two hundred furnaces are established, we have to consider that there would need to be a large organization behind it. A large complex like this would require an owner or owners, a board of directors, all the necessary factory managers, appointed superintendents, foremen, leading hands and so forth. There would also need to be a personnel department who oversaw the hiring or firing of workers and who decided which potential employees possessed the correct skills and crafts. They would need to organize working conditions such as hours, rates of pay, holidays, breaks and so forth. All this would have required numerous employees.

There would also be a need for Health and Safety Regulations, personnel who looked after the issue of protective clothing, and places, such as lockers, for the workers to stow their clothing and protect them from theft. All of this is right and proper for industry as we know it today, but the implications of the new discoveries are that these safety measures were already in place 4500 years ago – because it would not be possible to administer such a factory complex without them being already established and functioning.

In 1968, Dr Koriut Megurchian of the Soviet Union, unearthed what is considered to be the oldest large-scale metallurgical factory in the world situated at Medzamor in Armenia. The number of furnaces discovered – which amounted to over two hundred – produced between them a varied assortment of vases, knives, spearheads, rings, bracelets and even small items such as tweezers. This implies that these factories could produce spring steel – quite an accomplishment for those times. It was found that the workers wore all the necessary protective clothing that would have been essential to work in such an environment. These protective items also include mouth filters and gloves.

THE MODERN ANCIENTS

This equipment would have been issued to each man by the factory when they were taken on as an employee. Clearly, the complex's management knew of the dangers to health that existed; and we could go even further and suggest that the country may have had some kind of legal system in place, where, just as today, a worker could sue or take legal action against the management for respiratory or physical injury if necessary. One may further speculate that some kind of trade union system would have been in place to protect the interests of the workforce. Can we call our own era "advanced" when in the Victorian times (4400 years later!) workers were disregarded, exploited, and had to work in terrible conditions? Not to mention that they had no protection from ruthless employers, had to go to work as early as 6am after a crust of bread and a cup of water, and that children also had to endure these conditions and, for example, climb up chimneys in order to clean them. By contrast, it seems that the workers of ancient times had full protection from the obvious hazards of working in the midst of two hundred individual furnaces. They seem to have worked under a set of rules that offered some kind of protection – and were therefore likely to have had the protection of some kind of "factory act".

In these furnaces, copper, lead, zinc, iron, tin, manganese, fourteen different kinds of bronze, and even gold, was worked. These ancient smelters also produced a varied assortment of metallic paints, ceramics and even glass (so perhaps they had glazed windows 4500 years ago!). All of this implies that an organised sales, distribution, and exportation system was in place. Moreover, a transport system would have been needed as well as a large and organized mining operation worldwide to provide all the various ores. An efficient and well-established sea trade must have been established as copper and iron ore mines have been found all over the world. Some of these mines are unknown even to the indigenous natives – which shows how ancient they are. In addition, tin was extensively mined in Cornwall, much of which may have found its way to the factory mentioned in Armenia. It is amusing to envisage a workforce massive enough to man two hundred furnaces, all filing into work and perhaps operating under some kind of clocking-in system or being ticked off on arrival by a foreman. Such a well-organized complex may also have contained some kind of canteen where they could eat the equivalent of their chicken kiev's washed down with vodka. The more we learn about the skills and capabilities of the ancient peoples in these amazing discoveries, all of this seems likely.

As said, the mining of copper is ancient and goes back into the mists of time. Evidence of abandoned copper mines implies an ancient time of extraction; and such mines exist all over the world. The Greeks spoke of mines that even they considered ancient. The author Rene Noorbergen, in his book *Secrets of the Lost Races*, says "Beyond Europe, a number of recently re-excavated sites have already increased our knowledge with regard to pre-historic mining operations." Noorbergen relates that investigations conducted in 1967 and 1969 at Lion Cavern near Ngwenya in Swaziland, Southern Africa, showed that in pre-history, people had, "in some kind of organised effort, already mined out large deposits of hematite and speculative forms of iron ore".

On the Mediterranean island of Elba (known as having accommodated the banished Napoleon), there are also iron-ore mines whose origins are lost in pre-history. Hematite, an iron oxide, was in demand by the ancients as a form of cosmetic that was known as bloodstone and it was also used as a blood substitute in ancient burial ceremonies. Interestingly, hematite has also been found along with fossilised bone remains of the Neanderthal species – a species that expired 35,000 years ago. This is not necessarily to suggest that it was they who did the mining, they were likely not to have been advanced enough, so it is more likely that it was our anatomically modern predecessors (who had been around at that time already for thousands of years) who did the excavating. While Neanderthals were on the decline, it has been shown that the Cro-Magnon peoples acted and dressed with care; they did not look like the typical "caveman" as he is depicted in many cartoon-type drawings of today. This, if nothing else, gives us some idea of how ancient technology really is.

The implications of all this goes far. There would had to have been organised labour and all that goes with it. Experts in all fields relevant to the safety of the mining operations would have existed. Engineering methods regarding leverage, weight distribution, safety of the workers, the correct identification of weak points, and correct shoring-up of the dugout zones would had to have been in operation; the list of pre-requisites is quite lengthy. The correct route of the tunnelling operations – that would need to follow the ore seams – would be paramount to avoid wasted effort and to maximise the volume of ore that would be removed. And what of the ore stockpiles assembled?

The market for all this would likely have been in place as a functioning organisation; batches of the material would have to be designated as regards to their destination. Of course, strong transportation carts would have been needed to move it – sometimes to the coast to be put on ships and taken to overseas destinations. So this would have necessitated a large and well-organised communication and transport system between the buyers, accountants, logistics and administrators. This organisation would have required a complex recording system and numerous employees who were proficient in administration and mathematical skills. All of this would be little different to modern-day commerce, only on a smaller scale. Of course, it would not only apply to metallic ores. Items have been found in various archaeological digs that originated and were manufactured thousands of miles from their source.

It is clear that we are wrong in our estimations of the ancients of 4500 years ago only being involved in battles and warlike behaviour. All of that of course did happen, just as it does today, and the excavating and supply of all that metal would be just like the armaments industry of today supplying other countries with the material for weapons and war. As far as the ancient Greeks were concerned, their legends speak of Hephaestus (known as Vulcan in Roman mythology) who was expert in producing all

The Factory Act 2,500 B.C.

the necessary metals. The palace and workshop of Hephaestus was on Olympus and so he was known as "The Blacksmith of Olympus." Greek myths speak of the four ages of metallurgy: the first epoch being the Golden Age, the second the Silver Age, the third the Bronze Age, and lastly the Iron Age which we must assume reached its peak with the production of fine steel. We could ask, what put it into the metallurgists' heads that adding carbon to iron would give them the ability to produce fine steel?

Of course, the "big business" that existed in metal production – the existence of which is evidenced in the discovery of all those furnaces at Medzamor – was flourishing long before the ancient Greeks. And just as advancements in industry and technology take place at different stages in different areas of the globe (with some areas being less technologically advanced than others) so it was in ancient times.

It is known that the history of metallurgy began with soft metals and ended with hard iron. Copper is the essential ingredient of bronze. Just as is the case today, any new invention is assessed for its potential use in war. The ancients quickly realised that with their new discovery of bronze, and with its ability to produce far more durable and strong weapons for use in battle, that they would now become the prehistoric armaments industry. Copper mines flourished all over the world; copper was mined in Sinai, Crete, Cyprus, Spain and Portugal. The other essential ingredients for bronze came from equally widespread areas, such as tin, which was exported from Gaul, Spain and in large amounts from Cornwall.

As we have said, the transport of all this material obviously required a well-organised system. The alloying of metals and the various experiments that were surely needed, would have required long periods of trial and error in the same way that the ancient alchemists would have experimented in their quest to produce gold (if they ever did). The discovery of bronze at this time would have been a major breakthrough in their efforts and to arrive at the fact that bronze requires just one part of tin would have required lengthy experimentation over time.

It would have been realised, of course, that iron implements were heavy, brittle, and tend to corrode quite quickly in their attempt to turn back to what they were in the first place. Another factor we must consider is how efficient the primitive communications system was likely to have been – evidenced by the fact that advantageous discoveries seemed to spring up quickly, in faraway parts of the world, in quick succession.

With regard to the use of iron, it nevertheless still seems to have been used in manufacture in its raw state after the discovery of bronze – as evidenced by a pillar made of iron in Delhi that weighs six tons and is around seven and a half meters high! It has stood up to the tropical sunshine of India, plus all of the heavy rain and downpours in the area during the monsoon seasons for 1,500 years; yet it shows no sign of rust formation. Clearly this large artifact must have demanded superior forging skills. Some rather weak explanations of its rustproof capabilities have been offered – such as the idea that the natural oils in peoples' hands when approaching and touching it in awe have, over the centuries, rustproofed it. But this would not account for the higher part that people could not reach. It is doubtful that we could produce rustproof iron today. It therefore stands as a mute witness to the advanced skills of men who time has forgotten.

CHAPTER III

THE ANCIENT MINERS

We have mentioned the material 'hematite'; and the fact that this substance was used by the indigenous peoples in those places where it was found has been ascertained. And like other substances and practices, it has also been discovered in widespread places across the globe – for example in Tasmania off Southern Australia and at the tip of Tierra del Fuego off South America.

With regard to the ancient mines and the equally ancient miners, diggings that took place in 1972, not far from Ngwenya at Border Cave in South Africa and conducted by Adrian Boshier and Peter Beaumont, uncovered ten (now filled-in) mining pits – some of which were up to forty-five feet in depth. And again, hematite had been extracted here.

Just as today, any diggings deep into the Earth will uncover animal bones, but when human remains are found, everything comes to a halt for at least a while – depending on the age of the bones. It does not necessarily follow that bones uncovered have any association with ancient mining; a large number of people go missing every year without trace – their remains only turning up when, for example, the foundations for a housing estate are dug. However, in this case of ancient mining in South Africa, some artifacts were also found that may have been items of personal property; these items included finely-crafted knives that were still so sharp they could slice through a sheet of paper with ease.

With regard to the hematite extraction, it would seem that some kind of "overseer", or what used to be called a "tallyman", was busily recording the amount of material being extracted by making etchings on bone – which would have required some knowledge of mathematics. The care that was taken in this careful record-keeping indicates organisation rather than just haphazard digging and piling etc. And it shows that ore was an important product and must have had sufficient economic value to prompt the ancient diggers take track and carefully stack as they did.

Interestingly, some of the most fascinating evidence of pre-historic mining can be found in North America in the Keweenaw Peninsular and on the Isle Royale in Michigan. In the copper-rich Lakes Superior region, there are ancient mines whose origins are completely unknown even to the long-established local Indians. There are clear signs that several thousand tons of copper were removed at some early date in antiquity, yet not a single cultural artifact remains in order to assist us in ascertaining just who these miners were. The journal of the American Antiquarian Society stated "there does not seem to be any obvious indications of any kind of permanent settlement near to these mines. Not a vestige of dwelling nor a trace of human remains, not even a bone has been found!"

Yet more strangely, no trace of the extracted ore has been found for a distance of a thousand miles from the mines. To start with, these miners must have required breaks in their work routine; they must have sat and eaten some kind of food and drink. If the workforce was sizeable enough, and if the mining continued for as long as it must have done, then some evidence that humans had been there, such as discarded bones and other signs, would surely manifest itself. It is hard to imagine that the workforce simply came, worked all day, and then left without a break. Bones, clay bottles or drinking vessels (as they would have required liquid refreshment also) would surely have been found when even shallow excavation of the area took place; but there was nothing. But the strangest thing is, is that if no evidence of their excavated material was found within a thousand-mile radius, then this implies the unacceptable notion that they travelled a two thousand mile round-trip to their workplace and back.

As facetious as it may sound, the evidence from the Hindu texts is encouraging because of their very descriptive explanations. And if we take seriously the possibility that large and small vimanas were flying round in ancient India, then we have to ask the question, "did the owners of the mines hire vimana to transport their workforce?" After all,

The Ancient Miners

particularly if we think of the Medzamor furnaces in Armenia – they were in operation some 4500 years ago, the same time that the Hindu sagas were written.

The first discovery of the deep mine shafts in North America was made in 1848 by S. O. Knapp, an agent of the Minnesota Mining Company. When surveying the landscape and passing over the ground, he observed a continuous depression in the soil which, with his keen eye and lengthy experience in the field, he surmised was the disintegration of a geological vein. This led him to a cavern where he found evidence of artificial excavation. After clearing away the debris, he discovered a hole, together with tools and hammers, and at the bottom of the hole was a vein of ore that had not yet been mined out.

Perhaps the ancient miners considered this vein to be unviable for some reason, or uneconomical, and moved to richer pickings elsewhere. After this first find, S. O. Knapp went on to discover a second mine two and a half miles east of the Ontonagon River. Here, at a depth of eighteen feet, he discovered an already detached mass of copper that was found to weigh six tons. This mass was obviously made ready for transportation as it was raised up onto timber. It showed evidence of having been being hammered, or one might say "tailored", as though to suit the transport that would carry it. Presumably this large mass of copper would have to have been lifted at some later stage by a device we can only guess at.

Some of these prehistoric pits reached depths of sixty feet and some went through nine feet of solid rock in order to reach the veins. They must have possessed formidable tools. Obviously, before commencing the digging operation, there must have existed some very intelligent and experienced people who knew how to detect the veins and who must have been tasked with the job of deciding how economically viable any new mining site was likely to be.

It is generally believed that prehistoric and ancient peoples were shorter in stature than we are today, and had no fear of depth; yet, while many tunnels were found at considerable depths, one of these was found to be eight foot high. So why would these people dig a tunnel that was this high, with all the (otherwise unnecessary) material that would have had to have been extracted, if people only averaged around five feet high?

To accomplish all of these feats today would require advanced technology and sophisticated machinery – so this in itself is a good indication that these ancient people were far more advanced in their achievements than we have up to now assumed.

CHAPTER IV
THE ANCIENT DOCTORS

The established and instantly recognised symbol of the medical profession is the so-called "caduceus" of Hermes; it consists of a vertical staff with two snakes intertwined around it, topped with outstretched wings. Medical staff in the Armed Forces, for example, wear this symbol on their lapels, and it would seem that via this symbol, modern medicine acknowledges and pays respects to the sages of antiquity.

Significant examples of how advanced medical practices were in past history are found all over the world. These examples are also found in ancient dentistry where significant bridgework is often apparent. In one of the archaeological expeditions to the Valley of the Kings in Egypt, a number of ancient mummies were found. When investigating around the lower skull and jaw areas they found that many of the jaws had bridges and artificial teeth that had been quite professionally fitted and looked very much like the work of modern-day dentistry. Certainly, such clear evidence of ancient proficiency and skill in dentistry was an unexpected find, and it is not only in Egypt where such evidence has been found.

On the coast of Jaina in Campeche, Mexico, some Mayan skulls were unearthed. The jaw areas of these skulls, just like the Egyptian ones, show clear evidence of skilfully-applied dental surgery. These jaw areas show crowns and fillings that were still in place when they were found by archaeologists.

In order to last so long, these ancient dentists must have been quite proficient in their ability to produce very high-quality adhesives. As well as dental surgery, medical procedures of other kinds were also quite evident with regard to the skulls in general. Skilful surgeons of ancient pre-Incan times carried out surgical operations that even today would be viewed as quite intricate not to say risky with regard to the survival of the patient.

These delicate operations were carried out on the brain itself around 2,500 B.C., so some 4,500 years ago. And so this time period again seems quite significant and relevant to all the events we have discussed so far with regard to ancient technology, metallurgy and mining techniques. It is as though worldwide advancement blossomed into fruition all over the globe at that time.

This consideration supports the idea that the post-Diluvian Patriarchs and their descendants spread out from the mountains of Ararat to reinvigorate the Earth and to pass on as much of their knowledge as they could, before their own demise. Noah himself was alleged to have lived for centuries after emerging from the ark. We mention once again the discovery in Medzamor, Armenia, of the huge area of factories and furnaces which is only a few miles from the Ararat Region. So this sudden blossoming of knowledge could be considered a renaissance of pre-Diluvian knowledge and advancement – not only in metallurgy and warfare, but in all the branches of medicine and technology. And this renaissance was invigorated by the new teachers who were often referred to in many of the ancient legends as "gods".

More evidence of these aforementioned medical skills is to be found in Peru where clear evidence of "trepanning", or surgically removing a circular disc of bone from the skull to reach the brain, has been found. What is even more significant, is that these operations seem to have become so common that they appear in their thousands in the skulls of Peru. Evidently, quite successful trepanning must have been carried out over a long period of time. Surgical instruments were also found that had been skilfully manufactured to be purpose-made for the operations that they were to be used for. Some of these instruments were made of obsidian, and they included scalpels, bronze knives, pincers and needles for use in postoperative sutures.

It was also evident that many successful amputations had taken place. In the mind's eye, one can conjure up a scenario in a modern operating theatre, with caps, gowns, spotless surroundings, and a team of

The Ancient Doctors

specialists with high powered lighting and so-forth; so we have to wonder what kind of conditions the ancient surgeons worked under. But these advanced operations could not have been successful had they not at least had the benefit of clean surroundings, dressings, and antiseptics, otherwise many, if not all of the patients, would have died from sepsis, post-operation gangrene, or trauma and shock.

Evidently then, the ancient surgeons must have guarded against these effects. And it was indeed found that, to combat the dangers of surgery, these surgeons used gauze for dressings and apparently utilised cocaine, which was well known to the Incas, as an anaesthetic. The Incas were also familiar with, quinine and belladonna.

Long periods of experimentation must have occurred for this knowledge to be gained; it must, in part, have been gained by a natural and gradual process of building upon past knowledge and experience. An example of this occurs in the writings of Herodotus who described a policy that was adopted in ancient Babylon whereby the sufferers of various ailments were brought out into the street, and it was the moral duty of anyone passing by to enquire about the details of the patient's ailments and symptoms. If the passer-by had, in the past, experienced similar symptoms themselves, it was their duty to relate this to the sufferer and to advise on any known successful cures.

Assuming that the passer-by sufficiently explained how he or she gained relief for a particular ailment, the doctor, as well as the patient, would have learned much from this process. A great deal of effective treatments in herbal medicine and cures must have been learned in this way. To be sure, this pragmatic and simple ancient Babylonian method resulted in data that was gathered, logged and subsequently utilised. This formed the basis of pharmacopeia and diagnosis in years to come, and was certainly a common-sense approach when compared to the quackery that occurred leading up to and including the Victorian times. During the eighteenth and nineteenth centuries, many potions killed rather than cured, and the "bleeding" of patients – a procedure that was based on the assumption that the malady would pass out of the body with the blood – was commonplace.

Many of the drugs that we are familiar with and frequently utilise and rely upon today, such as penicillin, aureomycin and terramycin, have their origins in ancient Egypt. For example, a medical papyrus dating back to the fifth dynasty of Egypt, mentions a certain type of fungus that grew on still, or stagnant, water and this fungus was prescribed for the treatment of sores and open wounds; so was this a type of penicillin that was discovered 4,000 years before Fleming?

In fact, the ancients appear to have been well aware of the dangers of infection and seem to have invented various types of antibiotics for various different types of uses. We know that the ancient Greeks and the Chinese were aware of the need for antibiotics because, for example, they used warm soil and soya bean curd which was said to have antibiotic properties. They employed these substances to heal wounds and to eradicate boils and carbuncles. The ancient peoples of the Nile were also aware of the importance of hygiene, and the medicines utilised by them are now known to be far superior to the practices used by European doctors and surgeons of the Middle Ages; a massive amount of knowledge was evidently lost in the centuries between.

Amazingly, the Egyptian physicians of old were also just as advanced in their knowledge of the function of the heart, arteries, and the whole circulatory system – a knowledge that they gained through experimentation, dissection and learning. We know that they were also familiar, for example, with the important practice of counting the pulse.

Perhaps when we think again of the ancient Incan medical practices, we might hypothesize that they built advanced operating facilities and theatres. Perhaps we may yet discover the remains of such a facility in the ruins of one of the fine buildings that they constructed – or in the ruins of a building yet to be found. The architect of the Djoser pyramid, Imhotep, built in c.2,500 B.C., is considered by some writers on ancient Egypt to be the first recorded physician in history. And the date 2,500

The Ancient Doctors

B.C. is again sounding familiar being the date of the craftsmen of the ancient patriarchs, the date of the ancient Hindu texts and their fearsome weapons, and the date of the ancient foundries etc.

Ancient India also possessed advanced medical knowledge; doctors there were familiar with human metabolism, the circulatory system, and the nervous system. Vedic physicians understood which medical processes were necessary to counteract the effects of poison gas, and they also used anaesthetics, performed caesarean sections and carried out brain operations.

Amazingly, experimentations in the field of plastic surgery were also made by the ancients; in the fifth century B.C., the Indian physician Sushruta listed the diagnosis of 1120 diseases and described 121 surgical instruments. So again we wonder how these ancient peoples gained this knowledge and must conclude that it goes back far into the mists of time – a fact that negates the story of the first man Adam that persists in some circles to this day. Such advanced ancient knowledge and practice makes nonsense of the list of patriarchs that is so matter-of-factly related in Genesis. How can the alleged birth dates of the patriarchs that came after Adam be accurate when we consider that all the aforementioned knowledge – the advancements and practices – were already in place and functioning?

Another area of medicine in which the ancients came first is vaccination as evidenced, for example, in the *Sactya Grantham*, a Brahmin book compiled around 1500 B.C., that contains the following passage giving instructions on smallpox vaccination: "Take on the tip of a knife, the contents of the inflammation and inject it into the arm, mixing it with the blood. A fever will follow, but the Malady will pass very easily and will create no complications". So clearly ancient India has prior claim over Edward Jenner (1749–1823) as the inventor of the first vaccine.

We may think that we at least came first in the introduction of medical aid programmes supported by the state (in short, the NHS), but we would be wrong in this assumption also; the physicians in the land of the Pharaohs also received their remuneration from the Government and medical aid was free for all.

Those who study ancient Chinese history will know that the ancient Chinese Emperor Tsin-Chi, 259–210 B.C., had the ability to X-ray people and that he possessed a "magic mirror" which could illuminate the bones of the body. In the year 206 B.C., this "magic mirror" was located in the Palace of Hein-Yang in Shensi. If these dates are correct, it was installed there four years after the death of Tsin-Chi and must have provided a valuable service for many years. When a patient stood before this rectangular mirror, which measured 1.76m x 1.22m, the image was said to be reversed but all the organs and bones were visible – exactly as in the case of the fluoroscopes that we use today. The "mirror" was used for the same purpose – to diagnose disease.

It would be interesting to know how the dimensions of this mirror have been ascertained because it is doubtful that the metric system was being used in ancient China. It may be easier to interpret measurements from ancient Chinese writings than from the "cubit" of Biblical writings – but half a metre in length is considered to be a good estimate for the latter.

Why did we lose all this amazingly advanced technological and medical know-how? There must have been many scholars and wise ones in existence in those times (perhaps more so than today) so one would have thought that they would recognise the importance of passing it on through the generations to ensure that it was preserved.

It is said that a Chinese surgeon by the name of Hua Tuo carried out operations under anaesthetic over eighteen centuries ago. His practices in the chronicle of the *Hòu Hànshū* (or, *Book of the Later Han*), A.D. 25–220, reminds one of a logged report from a modern-day medical journal: "He first made the patient swallow hemp bubble powder mixed with wine and as soon as intoxication and unconsciousness took place, he made an incision in the belly or the back and cut out the morbid growth." Clearly, medical reports and accounts were made or we would not have this

The Ancient Doctors

information – so at least some of these writings were locked away for posterity. But it is unfortunate that subsequently so much destruction of these scrolls and documents took place – especially the destruction that we know occurred in the near and Middle East.

The above ancient Chinese medical report continues: "If the stomach or the bowel or the intestine was the part affected, he thoroughly cleaned these organs after having used the knife for the removal of the contaminated matter which had caused the infection, he would then stitch up the wound and apply a "marvellous" ointment, which caused it to heal within four or five days and within a month, the patient was completely restored to health." One wonders what these surgeons used for the sutures, but it is clear that 1,800 years ago in China, one would have stood a better chance of a successful operation and recovery than if one had taken ill in the dirty Dickensian streets of London.

The Lester Institute in Shanghai was founded by Henry Lester, a British trade magnate, in the 1930s. Here, instead of discarding and discouraging some of the strange organic items that were used as cures for various ailments since ancient times, the institute made an effort to establish a scientific basis for these ancient Chinese remedies. And indeed, from what we have learned from the proficiency of ancient Chinese doctors and surgeons, it is difficult to dismiss these cures as "quackery".

Many of the items and substances that were investigated in the Lester Institute were found by a Dr Bernhard Reed to possess a chemical reason for their effectiveness; this is despite the fact that some of these items sound very much like the kind of things that in the Victorian era counted as quackery. Who would imagine that treatments such as donkey skin, dogs' brains, sheeps' eyeballs, pig's liver or the many and varied seaweeds, might possess a scientific reason for their ability to cure? Yet, Dr Bernhard Reed found in each of them a scientific reason for their ability to cure. Surely a curious thing? Perhaps unknowingly, some of the equally preposterous items mentioned in Dickensian times and written off as quackery, may also have been effective but were not used simply because they seemed so ridiculous.

Another example of ancient medical knowledge existing long before it was thought to have been first discovered in Western medicine, is that of blood transfusion being carried out by the Australian Aborigines. This is a practice that apparently goes back thousands of years in Aboriginal culture – well before the early nineteenth century when the first successful blood transfusions were performed by a British obstetrician. The method used by the Aboriginal people involves the practitioner taking blood from the donor – either from a vein in the middle of the arm or one from the inner arm – by the method of sucking it out through a hollow reed. However, the exact technique, though shown to various investigative teams, remains unfathomable.

One would think that, if this practice lasted for thousands of years, it must have had a high survival rate; so how were the Aboriginal people so knowledgeable with regard to the various blood types that were only thought to have been discovered in the early twentieth century? In addition, how was a procedure such as this carried out hygienically in the hot and dusty conditions of the Australian outback?

It is entirely possible that the Australian Aborigines, together with the New Guinea tribesmen and the Indians of the interior of the Amazon, have a much more adaptable and robust metabolism than do people of European descent. Either way, it would seem that the Australian "medicine man" is in possession of ancient knowledge that has been passed down through very long periods of time. During blood transfusion, the medicine man is always completely aware of the right vein from which to take the blood, and yet more astoundingly, he gets it right when choosing an appropriate donor – which could otherwise only be ascertained if both patient and donor undertook a blood test.

One of the contexts in which Australian Aborigines perform blood transfusions, is in a practice whereby they utilize blood (presumably from the young) "to give vitality to the aged"! This is a practice that has probably been carried out ever since their "dream time" without ever having been written down.

Many of these ancient ways of doing medicine are gradually disappearing due to modern-day influences, but one practice that will no doubt be retained in one way or another is the knowledge and ability to practice birth control. The Australian Aborigines traditionally created oral contraceptives by collecting the resin from a particular plant and rolling this resin into pills for use by women. But, as said, the exact methods are almost forgotten as the young take less interest due to the pressures of modern-day influences. And so, just like in the case of the American Indians, a valuable heritage, gained over the centuries if not millennia, is gradually disappearing.

CHAPTER V

THE ARK OF MOSES

When the Hebrews, after their long dusty trek through the desert, finally reached Midian, or the base of Mount Sinai, their leader, Moses, was to ascend the mountain for a lengthy period of tuition. After a little drama about the question of what would happen to anyone else if they tried to follow Moses, he ascended the mountain to be tutored in the construction of a fearsome weapon and communication device that would be invaluable to his followers, that is, the Ark of the Covenant.

He would also be instructed on the building of a tabernacle in order to house the Ark and to accommodate the "other world creatures" known as the "angels", when they decided to come down from above. Moses was to be coached over a period of forty days in order to learn all about the idiosyncrasies of the Ark – the method of construction, the special wood to be used and so-forth, but most importantly, its power, purpose and communication ability.

The Ark was certainly a very special piece of kit that certainly proved itself useful, in times to come, against their enemy, the Philistines. The Ark was capable of many things. It was a fearsome weapon, it could augment destructive sound waves that could destroy any brick-built walls (with the help of the vibrating resonance of trumpets and the human voice), it could kill someone if they merely touched it (when selected to certain functions), and it could also seemingly control its own functions and switch itself on and off.

It would seem that gold was the necessary metal for its construction and also "shittim wood". Since Moses was already up in the mountain and no mention was made of him going down to obtain gold from the Hebrews or to find shittim wood, these materials must have already been up in the mountain in the possession of the "angels". As I stated in *Pillars of Fire*, I identify these "angels" as other-world beings, even if they were from what we call "heaven". They came from a place other than Earth

The Ark of Moses

and as such they were "extra-terrestrial". This tabernacle, built to house the finished Ark would be built far away from the prying eyes of the Hebrews. When one consults the Old Testament, it seems that Moses very much kept the secrets of the functioning of the Ark to himself. The Hebrews simply followed orders, and when the Ark was to be used for destructive action, they were informed only on a "need to know" basis. They assumed it was built only to house the stone slabs on which were inscribed the "holy" commandments on how humankind should behave.

This tabernacle to house the Ark was, it seems, not just a simple tent. The instructions with regard to its build take up many columns in Genesis. If this Ark still is in existence, hidden away somewhere, it would be the most priceless item on Earth. Not only because of its material value but also as a positive special weapon that would almost certainly be seized by the military of any country that it might be discovered in.

If it were simply viewed as a highly valuable item to possess, then it might have been simply melted down for gold bars, but what of the priceless items contained within? That is, the codes for human behaviour. We might imagine that if "other world" entities contacted certain gifted humans, then perhaps they felt they could only make open contact with humanity in general until they had seriously modified humanity's behaviour patterns. Yet the strange paradox exists that the Ark had deadly capabilities as a weapon and that this capability was then given to the Hebrews to efficiently kill their enemies. We could ask, why was it so important that Moses constructed the Ark himself? The other-world entities we know as "angels" could have constructed it themselves and then simply allowed Moses to climb the mount and receive his explicit instructions during his six-week course there. He brought the tablets down the mountain but not the Ark. Although King David was happy to utilise this special weapon in his many conflicts with his enemies he was rather "nervy" of it and at one point refused to even allow it into his Royal City. His son however, King Solomon, well established in history as the ultimate "wisest of men" thought he had discovered the secret of disarming the Ark via the removal of its carrying staves.

The Ark was supposed to be the protector, and the special weapon of the Hebrews, yet it acted independently at will. On two occasions mentioned in the Old Testament, the Ark killed people. On the first occasion it killed a Hebrew ally as he helped to steady it when it shook about on the cart that it was being transported on. On the second occasion, a certain "nosy parker" could not resist sliding the lid off the Ark in order to peek into it; he was supposed to be keeping it secure in his own home temporarily but he paid with his life and was found, when the rest of the band came to collect it, on the floor without his head or his hands.

Yet apparently, the only items inside the Ark were the tablets on which the codes for human behaviour were inscribed. These were not the original tablets, but a second set that had been provided by the "angels" because Moses had broken the original set. This had happened when he had thrown them down in a fit of rage when witnessing how the Hebrews had lapsed back to wicked ways while he was away. During his absence, the Hebrews would not have known whether Moses was alive or dead, and Moses had not told them that he would be gone for so long as he probably didn't know how long he would be gone for. As I said in *Pillars of Fire*, if all this really happened then the broken tablets would still be lying at the foot of Sinai covered in windblown sand. With regards to the Ark of the Covenant itself, it would be difficult to determine where to even start looking. But if a team can be assembled to climb Mount Ararat in search of the Ark of Noah, then it doesn't seem inconceivable that some latter day "Indiana Jones" may yet set out in search of the broken tablets. After all, to find these they would at least know where to start.

The second set of tablets, being housed in the Ark, could not have been put in a safer place when considering the Ark's formidable powers. The chosen ones who carried the Ark must have been well versed in the power of what they were transporting and knew that they must carry it at a formidable distance ahead of the masses. Later, it was housed in the

The Ark of Moses

specially constructed tabernacle, which again was located far ahead of the crowd to prevent one of them creeping up to have a "sneak peak" which no doubt would have been of great detriment to themselves.

If by some method it was discovered today, who would even dare to approach it? If such a team had been formulated, only a chosen group would know of it. It would be comprised of certain experts in their field such as physicists, radiation experts, biblical scholars, and specialists in other fields with top of the range detection instruments. This is on the assumption that were it found, it would still possess special powers to protect itself. Yet the strange thing is, that during the battles mentioned in the Old Testament, it was captured (or perhaps allowed itself to be) by the Philistines.

Perhaps it had the ability to record their conversations and learn of their war plans against the Israelites, but the Israelites were always at pains to find suitable accommodation for it. The most well-known deed depicting the Ark's incredible power was of course the destruction of the massive walls of Jericho, ably assisted by the trumpets and yelling of the Hebrews. All due, of course, to the scientifically proven destructive capabilities of sound. We mentioned that Solomon deduced that to disarm the Ark, the staves should be removed. He may well have studied the ancient writings of Moses and his clear warning, given by the "angels", that the staves should not be removed.

Moses was given this warning by the angels during his lengthy course of instruction up on the mountain. In the biblical Book of Kings it mentions the point when the Ark was installed in Solomon's temple, it states: "and they drew out the staves that the end of the staves was seen out in the Holy place before the "oracle" [the Ark?] and they were not seen without and there they are unto this day. There was nothing in the Ark, save the two tablets that were put there at Horeb [Sinai?]."

Regarding the various stories surrounding the mystical Ark, some of which are lost in the mists of time, we might mention the following account about the Queen of Sheba. We can hardly mention Solomon without bringing Sheba, the Queen of South Yemen and territories across the Red Sea in Ethiopia, into the picture. Her domain was quite extensive. The deeds, reputation, and alleged powers of King Solomon had spread far and wide – and this would be expected as all the gossip and tales worth mentioning would be related along the spice routes and silk roads by the camel trains that travelled them.

It was quite a formidable distance south from the domain of Solomon to South Yemen – traversing across such countries at Jordan and Saudi Arabia. The Queen of Sheba was in no desperate hurry to reach Israel and no doubt, there were frequent stops at the various water holes and oases along the way. When reaching the domain of Solomon who (as history has recorded) the Queen was enamoured with, as was he with her, they became lovers.

As a result of this union, a son named Menelik was born, who it is said became the first King of Ethiopia. This was to be expected as his mother's domain stretched across the Red Sea. And it was said that Solomon adored his son and that there was nothing that he would not give him if he asked.

At this point, Solomon was still in possession of the Ark of the Covenant which he housed in his temple. When Solomon asked his son Menelik what he could give him that was "not ordinary", he certainly did not expect the reply that he received – his son asked him for the Ark of the Covenant. Solomon, naturally quite shocked, replied "My people would not allow that, they would turn against me," but he was not deterred. With the proven wisdom of Solomon and the high intellect of his son, they eventually came up with a way to solve the problem, so Solomon said, "There is a way where you could take it, yet not take it."

The solution they came up with was that they would have a replica constructed – but this would remain a profound secret with only a few priests in Menelik's Kingdom knowing about it. Obviously, they could not trust one craftsman alone to make the entire replica, so it was made

The Ark of Moses

in parts by various artificers who would not be aware of the whole project – otherwise it would have been instantly recognised.

Nevertheless, there must have been some suspicion among the various craftsmen, particularly as they would be working with gold and the special wood that Moses had been instructed to use. So, Menelik slipped away with the Ark. No doubt his mother, the Queen of Sheba, was in on the plot but could nevertheless be trusted to keep it secret. And so Menelik traversed back along the route his mother was familiar with until he reached the ideal crossing point of the Red Sea into Ethiopia.

Menelik would install it in a special church that he probably designated himself. As a matter of fact, there is a church in Ethiopia that is said to be the final resting place of the Ark. The priests there are chosen in turn by the previous priest, and must live and die in the church – but this is a penalty they no doubt happily accept for the potential privilege of seeing the Ark, even if they can never tell anyone about it.

Of course, there are other stories that relate to the resting place of the Ark and eventually it becomes rather like the task of finding the genuine Camelot or the real court of King Arthur. One case that is widely considered, if not accepted, is that after the invasion of Ethiopia by Italian troops in the 1930s, the Italians captured it and installed it secretly in the Vatican.

However, most of the tales regarding the Ark are simply sheer speculation which will no doubt prevail until it is found – if that ever happens. But it has secured its position as one of the world's greatest mysteries. If the story of its theft from King Solomon is true, then he would no longer have been able to communicate with the "angels" or those on high, as it was also an oracle. We wonder whether the ability to do this was transferred to his son Menelik or not.

Solomon had had a great respect for the Ark and he would have remembered all the stories of his father's adventures with it during the battles with the Philistines. So after it was stolen, Solomon must have carried a great burden of guilt regarding the subterfuge carried out against him and in his allowing the Ark to leave his Kingdom. He had initially taken such care in reverently housing the Ark and making it secure in his temple inside which he had built a special enclosure to house it. As mentioned, during this time, Solomon had been able to communicate with the beings he accepted "gods" or "angels" and in earlier days, they had also descended into the temple itself.

Perhaps Solomon had constructed a special roof aperture through which the "gods" could ascend and descend and this is hinted at in the Book of Kings. It took Solomon seven years to build his temple and no doubt all kinds of special facilities were installed, some probably known only to Solomon himself.

When he was communicating with the beings "on high", Solomon obviously felt very secure with the fact that they were showing such interest in his Kingdom and in the land of Israel – and that they had done so since even before his father's time. He wanted reassurance that they would always be around and he questioned them about this via the oracle. The importance of the Ark and its capabilities were paramount but by the time his son took it, it seems that it had passed the peak of its power and former importance.

This was partly because the "gods" had made it clear to Solomon that he and his country would not always be their top priority. At one point Solomon tried to ascertain if the gods would always be around to call upon. In *Pillars of Fire*, I mentioned his communicating through the Ark or "oracle", by asking: "But will the gods indeed dwell on Earth? Behold the 'heaven of heavens' cannot contain thee, how much less in this house that I have built for you?"

It seems that by this point, King Solomon saw the Ark as less important because he knew that the gods would not always be there to be depended upon in the future. But we must therefore ask, how would King Menelik utilise it? Did he also use the oracle to communicate with the "gods"? We will never know.

The Ark of Moses

Is the Ark really in an Ethiopian church? Would a priest today accept that his only purpose in life is to live in that church and never leave it? Even if he had not laid eyes upon the Ark himself? Is the Ark lying there quietly humming and awaiting the time when it might communicate its most important message ever to humankind?

Modern theory has it that a top secret facility exists that contains objects of extreme mystery and importance from the past, but the obvious question is why? If such a facility containing these objects exists, then why not tell the world and put an end to all the conjecture? However, the Ethiopian church definitely does exist and the guardian priest has been observed moving about in the surrounding garden but never leaving it. There would be no point in a team of reporters descending on it for a story; the priests would simply disappear inside and they would learn nothing. Yet, this has more going for it than the idea of a secret "Hangar 18" or storage facility based on some kind of Hollywood "Indiana Jones" story.

CHAPTER VI

MAKE ME SOME GOLD

Chemistry was known in ancient times as "alchemy". It is one of the most, if not the most, ancient of skills and many of the ancients dabbled in it – some claiming success. Those who claimed success, stated that they had transmuted base metals into gold – and much historical documentation exists to support their claims.

The general scientific thought that prevailed for a long period of time was that chemical elements are all quite stable and cannot be altered in any way. So, anyone who claimed that they could alter chemical elements was branded a charlatan and put their own life at risk into the bargain.

But of course, science had to "listen up" when the highly renowned physicist, Ernest Rutherford, actually transmuted nitrogen into oxygen and hydrogen in a process that required bombarding it with helium. Of course, this turned scientific thought back to the subject of whether there might have been some truth in the alchemists' claims; they had been a thorn in the side of scientific thought for so long that it was irritating to have to again consider their viability.

There are a large number of medieval manuscripts that go to great lengths to describe the various types of equipment that is required in alchemical practice. Perhaps we should stop here and ask why they were ostensibly being so free and easy with all this information if it was ostensibly such a secretive practice; perhaps we should consider that this information was in fact a purposeful diversion, "misinformation", that was put out there to put those unsolicited individuals who sought this secret knowledge off track. If would-be alchemists starting making gold in large quantities, then the true value of the gold made by the genuine practitioners would drop considerably – which is another point that tends to reinforce the assumption that much of the detail freely given may have been worthless.

Nevertheless, much detail was laid out with regard to all the necessary items required in order to practice "the great work" as it was known. There was a need for retorts, glass vessels, various types of distilling equipment and stills, heating equipment such as furnaces and much other equipment besides – all of which makes it clear that this was not a poor man's hobby. One would have to be in possession of sufficient funds to prepare oneself to commence the operation and a reasonable amount of gold would have to be produced just to cover one's initial outlay, but obviously many were not deterred and most likely had the wealth to begin with.

It would not seem likely that one would go to all this effort and expense if it were unlikely to produce any positive results. No doubt, the would-be practitioners were spurred on by the fact that for centuries before them, the alchemists positively claimed that they could transmute mercury, tin or led into gold.

The study of many ancient and historical documents makes it clear that there was a general fear of the alchemists by certain rulers – and with good reason. The rulers recognized that anyone who could produce gold would be a threat to the state, and to the economic stability of that state, because the production of gold would certainly have affected any gold standard that existed.

However, although this fear of alchemists existed among many kings and rulers, others positively encouraged it. For example, King Louis XV of France employed the services of one of the greatest alchemists who ever existed in an attempt to boost his own wealth and fill up his coffers. In *Strange Realities* I cover the story of the great Count of St Germain who was a master alchemist and producer of the elixir of youth. He also claimed to be able to manufacture diamonds. King Louis saw him as a godsend rather than a threat, and even made it clear to St. Germain that he wished to learn the "great work" and become a pupil of his himself.

By contrast, in the year A.D. 300, the Roman Emperor Diocletian issued an edict in Egypt in which he demanded that all of the available books that described the process of making gold and silver were to be collected and thrown into the flames.

So this type of activity, and the general worry that surrounded the effects on royal and state economies, proves to some extent that the art of alchemy was indeed taking place. It seems that the primary concern, rather than the more communal concerns raised above, was that the practitioners would themselves become excessively wealthy – and that since money is most assuredly "power" that this would "buy" officials by (in modern parlance) making them offers they could not refuse.

In addition, they would have the wealth to buy territories. Diocletian wanted all or any written art works and documents plus any records concerning the ancient art withdrawn from circulation and disposed of. This, of course, would have generated an enormous effort to copy as much data as possible before handing over the originals to prove they were complying with the decree. So in reality, such a decree would only have forced the real practitioners underground.

With regard to the power that great wealth affords, on record to review, was the case of the praetorian guard, Didius Marcus, who was a multi-millionaire. He was alleged to have bought the whole Roman Empire for the equivalent of thirty-five million dollars. Nevertheless, in spite of his great wealth it did not prevent him from being beheaded by the Emperor Septimus – proving that another kind of power still reigned – that of the sword.

According to the alchemist Zosimus, who lived in A.D. 300, the Temple of Ptah in Memphis, Egypt had all the necessary equipment and furnaces to enable the practice of alchemy and Zosimus was referred to as the god of the alchemists and its patron. And such a known alchemical laboratory, unless it was disguised as a normal industrial facility for weapons manufacture for example, would had to have been extremely well guarded rather like a Fort Knox of ancient times.

Over the centuries, other alchemists, each in their turn, took the crown of being the most renowned. In the eighth century, the Arab Jabir ibn Hayyan systematized all the alchemical knowledge available from Egyptian sources and subsequently became known as the father of this ancient practice. He was a functioning alchemist himself and he was one of those that took great pains to describe all the special equipment and apparatus that would be required when originally setting up a laboratory.

Jabir even went so far as to describe the necessary prerequisite moral and mental characteristics of anyone wanting to take up apprenticeship of the art. These characteristics sound a bit like the necessary attributes that we mentioned were laid out in the Hindu texts on the ideal characteristics for a pilot of a vimana. However, Jabir also goes on to say that the artificer ought to be well-skilled and perfected in the sciences of natural philosophy. He assured the students that "copper could be changed into gold and by our artifice we can easily make silver"; so, he was certainly confident and convinced of the truth of alchemical practices.

It would be quite easy to dismiss all of this and the other data were it not for the fact that he has secured his place in the history of modern chemistry. We could also take the same view with regard to the aforementioned Count of St. Germain. The art and frequently-mentioned practice of alchemy was not restricted to just a few countries; it was allegedly known in ancient China at least as early as 133 B.C., and much later it was still practiced in China. The biography of Chang Tao Ling, or Zhang Daoling, A.D. 34–156, who is stated as having studied at the renowned Taixue – the Imperial Academy – makes references to the "treatise of the elixir refined in nine cauldrons" that he found in a cavern – and the author of the treatise was said to have been the Yellow Emperor of 2600 B.C.

It is specifically stated that the basic but primary ingredient of Chinese alchemy was cinnabar, or mercury sulphide, and that this was to be used in the process of transmutation and was also necessary in the procedure of compiling the "elixir of youth." We have mentioned the Count de St. Germain – he was thought to have discovered this profound "secret of eternal youth," and not just through embellished legend. Historical documents refer to people who knew the mysterious Count over several decades and who testified that he always appeared to look young and did not age through the decades that they knew him.

It is well worth reading about this amazing character in my book *Strange Realities*. The Count actually studied in China and could speak the language. The historical manuscript, *Shih Chi*, that was written in the first century B.C., clearly states "you may transmute cinnabar into pure gold."

As in other countries, the reality of, or at least the practice of the alchemical arts, is somewhat verified by the fact that a law was enacted in China in 175 B.C. that clearly warned against the practice of "counterfeiting gold by alchemical means." Obviously, the production and the output of so-called "counterfeit" gold must have been a considerable worry in China as well as in all the other countries where similar warnings were issued.

Perhaps the term that was used in these accounts – the term "counterfeit" – is incorrect. "Counterfeit" implies "false" or "phony", but the gold that the various experts in the field produced (and we could include the Count St Germain among them) was said to be as pure as any other gold in use.

It must be significant that the various elements that are mentioned as necessary in the production of alchemical gold, repeat themselves across the worldwide legends involving alchemy. And it is no surprise that the transmutation of these elements is still a puzzle to modern-day physicists.

A book written about fifty years ago by the author Andrew Tomas, deals with alchemical practices and mentions how alchemy was a worldwide phenomenon. Tomas writes, "The Hindu expositors of the art thought that mercury and sulphur were the primary elements to use but unlike the Chinese and European alchemists they attributed positive polarity to mercury and negative to sulphur, they were also convinced of the reality of the elixir of life and immortality as well as the practice of gold making." In view of the fact that the art of transmutation and the production of gold placed its adepts in a dangerous and quite vulnerable position (simply because of natural human envy, malice and possible robbery that might lead to loss of life and suspicion from the state), they had to be very secretive about their operations and had to select the locations of their laboratories very carefully.

The profits and gains to be made must have outweighed the risks and made it all worthwhile. Nevertheless, the practitioners also used charts and enigmatic texts with a variety of special symbols that would only be recognised for what they were (and their relevance understood) by other pursuers of the art.

The importance with which many European countries regarded this secret art is indicated by that fact that some countries, for example during the time of the Spanish Inquisition, employed "enforcers" who zealously tracked down and liquidated anyone who was accused or even suspected of practicing the so-called "magical sciences of the heathen eastern countries."

However, for all the mass of information and data from all the worldwide sources appertaining to the secret art, the question of whether genuine gold has been produced is, surprisingly, still not hotly debated – primarily because no scientifically provable and methodical process seems to exist. It is quite surprising therefore, that it has stood the test of time through all the years of wonderment and conjecture. Nevertheless, certain decrees and documents from many countries make it quite clear that the various authorities had little doubt that the transmutation of metals would at some point become a real possibility.

In *Strange Realities*, I gave an account of the famous alchemist, the Count of St Germain, and how Louise XV evidently had no doubts about the Count's proficiency in the art and indeed, moreover, reaped the benefits of it. Since the King was a real historical figure, this must be one of the best pieces of evidence in favour of the reality and validity of the practice. And if all this gold production really did take place, then much of it may be still in existence.

During the thirteenth and fourteenth centuries, alchemy must have been fairly widespread as it attracted the notice of the Vatican. The practice of the science was forbidden by a papal bull issued by Pope John XXII in the year 1317. This document was titled "Spondent Pariter" and it condemned any alchemists that could be found, to exile, and also established very heavy fines against them. All these widespread prohibitions of alchemy are somewhat bewildering when we consider the equally widespread denial of it – not so much with regard to its existence but with regard to the practitioners actually being able to produce pure quality gold.

In his book *We are not the First*, the previously-mentioned author Andrew Tomas made quite an interesting point, he said, "A no smoking sign, put up in a public transport vehicle is there because people have and use tobacco. Therefore, what was the reason for all those 'no gold making' orders if there were no legal cases of illegal transmutation in the first place? It surely would not have been worthwhile wasting quite expensive parchment on long sternly worded decrees if there was nothing to warn about". King Henry IV of England also took action against the practitioners of alchemy by issuing an Act in 1404 declaring that "the multiplying of metals was a serious crime against the crown." Obviously Henry IV was not as crafty as the later Louis XV of France or he might also have utilised, rather than ostracised, the best professional alchemist that he could find in order to make sure that the Crown got its share.

This is especially the case when we consider the political climate in Europe at the time; the edict was issued by King Henry IV during the time of the Hundred Years War (1337–1453) and the Peasants' Revolt – so one would have thought that he could have utilized all the gold he could possibly obtain for armaments as well as for his own coffers.

It has been suggested, in this regard, that it would seem unlikely that a king of England would sign a decree against a "mythical and unproven" practice while involved in the process of waging a real war in France and fighting angry serfs in his own territory. Others might say "Here is yet more proof of the reality of alchemy." Yet surely if there were that many alchemists, the king could have rounded them all up and persuaded them to pursue their art under Royal contract – and then proceeded to levy a substantial tax on them so that they could partly or perhaps fully finance the war with their gold (and gold of course, like money, "talks in any language").

In any event, by the fifteenth century it was clear that large amounts of gold were appearing from some unknown source and that was an established fact. Strangely, to use modern parlance, a complete U-turn was apparent in later years regarding the legality of producing gold. During the reign of King Henry VI, around the year 1444, it is written that he not only tolerated the mystical art but granted permits to a couple of gentlemen known as John Cobbe and John Mistelden in order that they might practice the "philosophical art of the conversion of metals." Also, amazingly, these "licenses to operate" were duly approved by parliament.

All this alchemically-produced gold was not only produced in the form of ingots, but it was produced as coinage as well. So all this swapping and changing of the legislation makes it clear that the old adage is quite sound – "if you cannot beat them then join them," or in this case employ them.

So the Crown eventually became quite tolerant of the practice as long as it received its due share for his majesty's mint. It seems that at last, the royalty came to their senses by taking this pragmatic approach, but how long would it last? As we have suggested, perhaps there are large amounts of alchemically-produced gold in the Royal Mint, or in Fort Knox, or in who knows how many other countries – and if it is as pure as claimed by the practitioners in ancient writings, then it would be no less valuable.

So, more significant than Henry IV's royal ban on alchemy in 1404, was the ban's official and complete repeal in 1688 by their Royal Highnesses William and Mary of England. The edict reads as follows:

"and whereas, since the making of the said statute, divers ["divers" in Old English "divers" meant "various"] persons have by their study, their industry and their learning arrived to great skill and perfection in the art of melting and refining of metals and otherwise improving and also multiplying them."

Obviously, the repeal of the ban was a sensible approach as it kept the alchemists "at home" as it were, thereby assuring that the Royal Mint "got its cut." And the Act of Repeal explicitly makes it clear that this was its purpose – to discourage them from leaving the country. The text of the Act states that from the reign of Henry IV onward, many Englishmen (the alchemists) went abroad to foreign lands in order to exercise their art which was of great detriment to the Kingdom – and was, in short, a form of tax evasion and something that still goes on of course.

The aforesaid decree announced that all the gold and silver that was to be extracted by application of the aforesaid art should and must be turned over to their Majesty's Mint, located in the Tower of London. The precious metal would then be purchased by the Crown at the full market price.

This was fair enough and obviously a good deal for the alchemists – but to keep on the move and be one step ahead of their enemies, and of the "revenue men", must have been a great inconvenience for them. This was an inconvenience especially when considering the large amounts of equipment that they would have had to transport about and that was listed as being essential for the art, not to mention the furnaces, especially if they had to transport all this abroad.

It would seem that William and Mary were very liberal minded as after this change of policy allowing the practice of alchemy, the King and Queen even made a declaration that indicated or encouraged the desirability of actually studying the art. Clearly, the sovereigns of England at that time realised the advantages of controlling the gold reserves and hence the economy, but if too many were producing it, then it may begin to have the opposite effect. It is a fact of human nature, ingrained in most humans ever since childhood, that when something is forbidden, it seems to make them want it more. For example, before the Act of Prohibition was repealed, there was a massive rise in alcohol consumption, even by those who could formerly take it or leave it. And in the United States of course, it enriched the criminal gangs.

In his book *We are Not the First*, Andrew Tomas claimed that he had actually examined a specimen of alchemically-produced gold. It was installed in the department of coins and medals situated in the British Museum in London. The artifact was in the form of a ballistic item, that is, a bullet, which is apparently the form that it originally took before it went through the transmutation process. Apparently, the register in the British Museum contains the following brief entry: "Gold made by an alchemist from a leaden bullet." It is said that it was transmuted in the presence of Colonel MacDonald and Dr Colquhoun at Bupora (possibly India) in the month of October 1814. Tomas made it clear that although detailed information regarding this event is lacking, it is important to note that the presence of the object and its acceptance by one of the foremost museums in the world gives weight to the belief in the practice of alchemy. One might cynically suggest that it may have been more advantageous to transmute a cannon ball rather than a bullet – but perhaps weight and size are factors to consider in the transmutation process.

Johann Helvetius (1625–1709) was physician to the Prince of Orange and was also known to be a practitioner of the art of alchemy. A person named Porelius, who was an Inspector General in the mines of Holland, came on a visit to the laboratory of Helvetius in order to observe his alchemical work. After having obtained a sample of gold from Helvetius, he went to an established jeweller, named Brechtel, in order to have him carry out a full analysis of the sample. It is said that after rigorous testing, the gold was found to be pure.

We have used the term "transmutation" quite frequently here but what exactly is it? The process of transmutation can be illustrated by the example of the element plutonium. Plutonium is an element that does not exist on Earth, yet it can be created by a process of nuclear physics. This process could be called transmutation, but with most items, or elements, that are frequently mentioned in accounts of alchemical transmutation – such as with the transmutation of mercury into gold – a change in the element's atomic number, it's make-up, is required.

The number of electrons, their orbits and the organisation of protons etc, determines the make-up of the elements. As stated, elements such as mercury, lead, and so-forth, are frequently mentioned in alchemical documents – and it is noteworthy to realise that in the periodic table of elements, the atomic numbers of gold, lead, and mercury are quite close together (i.e. gold is 79, mercury 80, and lead 82). Indeed, one might call them neighbours.

Regarding the discovery of the table elements by the Russian Ivanovich Mendeleev, it was said in a recent television programme that all his knowledge about the periodic table came to him in a dream. This is not disputed but it is nevertheless very hard to believe. Many people have experiences during the day that cause them to say "that broke my dream", that is, the experience reminded them of a dream they recently had. However, to have a dream and then write down such an extensive account of all the data and various elements and their atomic structures etc., is hardly feasible. It would have to have been telepathically imprinted into those memory processes within the brain that remember everything and forget nothing. Usually, one cannot conjure it all up by instant recall, although hypnotic regression has been shown to do so. But then we might ask ourselves, what "force" imprinted this information into the brain of Mendeleev? But that is the subject of another story.

Nevertheless, indisputably, in 1879, Mendeleev was the first individual who formulated a table of the elements and then arranged them in accordance with their order of weight – weight that increases in accordance with their atomic structure. It is hardly possible to imagine that the ancient alchemists were actually aware of all this scientific detail, let alone able to manipulate the elements to cause any change in their composition, but what alternative explanation is there?

It is said that the ancient science of alchemy was first introduced to Western European circles in the eighth and eleventh centuries by Arab practitioners at a time when the ruling elites were far less tolerant of the practice of alchemy. In fact, many rulers were quite hostile in their attitude to alchemy and this obviously required the alchemists to be "ready to march", as it were, at any time. As we have said, it was of crucial importance that practitioners had the ability to "pack up and go" at a moment's notice as their lives may have depended on it.

As a result, the alchemists drifted from place to place and indeed from country to country, but they would always have had to be very careful in their choice of area in order to safely set up their laboratory again. They would never, for example, have discussed their art openly in general conversations except with other alchemists. It is written that some sovereigns subjected certain talkative men to torture in order to obtain the alchemist's formulas.

In *The Compound of Alchemy* written by Sir George Ripley in 1471, he strongly advised student practitioners to keep quiet about their art and to keep "secrets in store unto thyself." The very wealthy, and the nobles and royals, were always on the lookout for, and seeking information about, the practitioners for two main reasons – either to exterminate them or to exploit them. Another notable – the great theologian, friar, philosopher and scientist, Albertus Magnus (1206-1280) – gave the following advice to his students: "to carefully avoid association with princes and nobles and to cultivate discretion and silences." Yet, if you read my book *Strange Realities* you will find that one of the greatest practitioners of the art of alchemy (if not the greatest) disregarded this kind of advice and in fact did exactly the opposite – often rubbing shoulders with kings, queens and nobles such as King Louis XV of France. I refer to the enigmatic Count of St Germain.

THE MODERN ANCIENTS

The great Dr Paracelsus (1493–1541) was said to have discovered zinc and was also the first person to identify the element of hydrogen. And so once again, it is tempting to imagine that these great thinkers might have discovered the table of elements, or at least made an attempt to compile one, long before its discovery by Mendeleev in the late nineteenth century. The name and fame of Paracelsus as an alchemist spread far and wide during his lifetime, so much so that it is said that his tomb, located in Salzburg, was opened up by desecrators who were obviously looking for any alchemical secrets or treasure that may have been buried with him. Although it was said that nothing was found in the coffin, apparently the famous sword buried with him and rumoured to house the mystical "Philosophers Stone" in its hilt, went missing after this time. It should be noted that the reference to "stone" in this context could be misleading because it may have actually referred to a "policy" or "creed" rather than "stone". After all, what use would a "stone" have been? It would have provided no obvious insight into the secrets and processes of alchemy.

Nicholas Flamel (1330–1418) was another notable alchemist. At the risk of repetition, it is hard to imagine that a subject and profession involving so many high-profile characters was nothing but nonsense. On the one hand, when we think of all the things that an alchemist such as Count de St Germain was supposedly capable of, it seems like fiction. But then, when we remember the high-profile names such as King Louis XV and the written accounts found in the memoirs of the nobility – then it seems to prove the validity of these alchemical feats. It is claimed that Nicholas Flamel came across an ancient book titled *Abraham the Jew*. From the title one might assume that it contained stories from, or referenced, the Old Testament, but this book was written in an unknown language. However, after intensive analysis it was determined that the data within was relative to the subject of alchemy and to be sure, Flamel would be well qualified to determine that supposition.

Flamel studied it intensely and allegedly, by the use of its instructions, he first transmuted half a pound of mercury into pure silver, and then a few months later on 17 January 1387, he used the book to produce alchemical gold. Again, the preciseness of these dates, names and so forth, seems to add validity to the idea that the alchemical process at least sometimes actually did do what it claimed to do.

Flamel was not an avaricious man. It is clear from the writings about him that he was in fact quite the reverse and that he used the gold for good and positive deeds. It was said that he built many churches and also hospitals. A lot of this was achieved in Paris during his thirty-six years of practicing alchemy there; his practice was evidently very profitable and indicates just how voluminous the results of alchemy and the creation of what we could call "gold bullion", could be. His local banker would clearly have been very pleased with this particular client, but as said, Flamel was not driven by the growth of his personal wealth.

Flamel is recorded as having stated: "In the year 1415 after the transition [one supposes it would be natural for him to use an alchemical term] of my faithful companion whom I shall miss for the rest of my life, she and I had already founded and endowed fourteen hospitals in this city of Paris as well as three completely new chapels, decorated with many handsome gifts, together with good incomes." This last phrase implies that he had also allowed for later upkeep and repairs, roof leaks etc. In addition, when he said that he "founded and endowed" with regard to the churches, he would have meant that he also provided all the seats, pews, internal décor and glazing etc. He also mentions that seven other churches had undergone many general repairs and improvements to their cemeteries – all achieved via alchemical gold.

Flamel also mentioned similar work that he and his wife had carried out in Boulogne. Obviously, all of this must have cost an enormous amount of money. He also wrote that on some of his churches "he caused to be depicted many marks and signs from the book of *Abraham the Jew*." Clearly, Flamel owed a lot to this mysterious book.

It is said that these signs and marks could be seen two hundred years ago, in places such as the Cimetière des Innocents, the church of St Jacques de la Boucherie, and the church of St Nicholas de Champs, so why are these signs not still visible today? Of course, buildings decay, sometimes they are destroyed, cemeteries get neglected, built on, or paved over and so forth, so perhaps there is a natural explanation. It must be said that the Old Testament, when describing the life of Abraham, makes no mention of any alchemical skills that Abraham may have had. It might be easy to assume that Flamel's reference to that name referred to the biblical Abraham, but Abraham is a relatively common name in the Jewish annals. The biblical Abraham was largely preoccupied with his extensive herds and flocks of cattle, before his "calling" from those "on high".

So Flamel owed a lot to *Abraham the Jew*, as did the ecclesiastical fraternity in France who gained from his generous donations. *Abraham the Jew* was not a work of fiction any more than alchemy itself was fictitious; the book was listed in the *Catalogus Librorum Philosophorum Hermeticorum* which in spite of all the words was a single document or book. It was issued by Dr Pierre Borelli in 1654. According to Dr Borelli, Cardinal Richelieu ordered a search for alchemical works in Flamel's house and in all the churches that he had built. This was not an act of aggression against alchemy on the cardinal's part, but was instead a search for knowledge and a wish to prove the whole issue of making gold which was always surrounded with such an air of secrecy.

The operation, set up by Cardinal Richelieu, was obviously a success, as the Cardinal himself was actually observed reading the said book. But did he understand it? Flamel himself had to study it very intently before utilising its alchemical data. During this period of study by Flamel he must have been intrigued or perhaps even perplexed at certain points because he made many notes in the margins. We know that at least the Cardinal did not immediately consign it to the flames as other ecclesiastical figures had done with other important writings – such as in the case of Bishop Diego de Landa with regard to the ancient Mayan documents. Perhaps Richelieu preserved the book because he was aware of all the good that had come of it via the work of Nicholas Flamel.

We mentioned George Ripley earlier. He was a known English alchemist of the fifteenth century and as an individual he was equally as memorable as Flamel. Elias Ashmole (a scholar of the seventeenth century and namesake of the Ashmolean Museum in Oxford) mentions George Ripley a number of times in his poetic writings. And another writer of the seventeenth century, Thomas Fuller, also wrote about Ripley in his book *Worthies of England*. He relayed a story here that an "English gentleman of good credit" when travelling in Malta, saw a record there that revealed that Ripley gave £100,000 per year to the knights of Rhodes and the Order of St John of Jerusalem. This money was to be used to help Rhodes fight the Turks. Most certainly £100,000 was an immense sum of money in those times, but as we have mentioned with regard to the other alchemists, huge amounts of gold must have been made to make this kind of money. And again we think of all the vaults and bank deposits that must have been needed to store it all, and to all the deposits that are currently in facilities such as Fort Knox and that may, unknown to themselves, contain a mass of alchemically-produced gold

And other alchemists were making so much gold that one of them offered to finance the crusades, and another to clear the national debt of his country. So we have to ask ourselves, where are all the alchemists of today? Why does any country in modern times have national debt? Did the secret art just fade away with no apprentices taking it up? And how could this have happened if it was so lucrative? To lose such knowledge doesn't seem to make sense. It was recently stated that due to the coronavirus pandemic and the response to it – the slowing down of commerce and industrial output etc – that many countries are in debt to the tune of many millions of pounds or more. Do we have any alchemists available to solve the problem?!

THE MODERN ANCIENTS

These questions sound flippant but they are nevertheless pertinent. We could also ask ourselves why all the ancient and effective cures and medicines disappeared so that in Victorian times people were treating themselves with leaches. What happened to the super-powered vimanas so that we had to start all over again with Wilbur and Orville Wright? We could continue but it seems to be a feature of human technological advancement that over the centuries advancement and forgetting rises and falls rather like the peaks and troughs on an electronic "sine wave". It is a matter of record that Pope John XXII, who originally issued a papal bull against the alchemist's and their practices, actually developed an interest in the subject himself. And there are a series of incidences like this throughout the long history of alchemy in which the authorities first condemn, and then capture or control the practitioners and the science.

In the case of Pope John XXII, it is possible that having perused so many of the documents on alchemy that he himself confiscated, he decided to study and to actively experiment in the science of transmutation himself. In fact, he even wrote an alchemical work on the transmutation of metals and in this work he admitted that he had worked with those familiar with the art. And again he uses the strange phrase, the "Philosopher's Stone", writing that he "worked with the Philosophers Stone in Avignon". But this reference to "stone" does not seem likely to be a literal reference. Most scholars would go with the explanation that "stone" actually refers to the "code" or "philosophy" that alchemists must abide by. So if Pope John XXII worked with the "Philosophers Stone" then it meant that he worked in accordance with the "Philosopher's Code of Practice".

In his laboratory in Avignon, it is written that he alchemically manufactured two hundred bars of gold, each weighing one quintal (or 100 kilograms) – quite a formidable amount of gold! And if he produced this amount of gold, he likely also produced coinage. And sure enough, after his death in 1334, 25,000,000 florins of gold were discovered in his treasure vault. We wonder, might this treasure vault still exist? One would imagine that more than one vault would be necessary to accommodate that much gold. Is there more gold in the Vatican's vaults than there is in Fort Knox? And if so, is it put to charitable causes?

For all the documentation and data that seems to point to the authenticity of alchemical practice, there are still doubts surrounding it, in particular in the volume of bullion that it may or may not have produced. There is said to be a museum in Vienna that contains compelling evidence of the practice of alchemy in past centuries. This evidence comes in the form of, and is catalogued as, an "Alchimistiches Medaillon". It is an oval-shaped medal measuring 40 x 37cm in size, and weighs seven kilos. Except for the upper portion which comprises around one third of the disc and is made of silver, the rest of the medal is made of solid gold. One might suspect that the medal is only gold-plated, but tests have proved that its composition is pure gold. There are a number of notches on the medal that were caused during its tests for authenticity – and we will explain these later. There is quite an interesting story behind this object and how it came to be produced.

The reticence to believe in the validity of alchemy is not necessarily surprising because some accounts, through their outlandishness, reinforce the idea of alchemy as mere fantasy. One such account relates directly to the above-mentioned medallion and how it came into existence. This account corroborates the basic story but seems to be embellished for dramatic purposes. Andrew Tomas also reminds us that the more ridiculous-sounding and highly embellished stories may have been deliberately made to sound far-fetched in order to divert attention away from the practitioners – so that they could stop "looking over their shoulders". If the stories seemed too ridiculous to be believed, then the authorities might leave them alone and stop trying to track them down. The story of how the medallion came into existence fits such a description of outlandishness. However, the creation of the medallion and the evidence of its authenticity, through the tests that it went through, is the story's saving grace.

THE MODERN ANCIENTS

In the sixteenth century in an Augustine monastery in Austria, there lived a young monk whose name was Wenzel Seiler. He could not be described as being totally committed to the calling and missed the more worldly activities of life. In fact, he was bored with the routine of the cloistered life and began to feel trapped there. Although Seiler's background was aristocratic, he was not wealthy and was without financial resources, so he could see no way of getting out of the monastery. However, he had one hope in the form of an older friar who befriended Wenzel and told him that there was treasure buried somewhere in the grounds of the monastery but he did now know where. So, between them, Wenzel and the friar tried to figure out where it was most likely to be buried.

After almost abandoning their search, they decided that the most likely location of the treasure was underneath a column. We pause here and contemplate why anyone would wish to bury anything "under a column" – it would surely be a difficult location to bury anything and might risk collapsing the load-bearing structure. Perhaps it was buried there before the abbey was constructed and the column erected. Nevertheless, after a lengthy search (still leaving us wondering why Wenzel would even imagine it to be under a column) they did indeed locate the treasure. It was found in an old copper chest. However, when they looked inside expecting to find gold and silver, they found instead an old parchment with strange signs and symbols on it and letters written in a language that they did not recognise. And along with the parchment, the chest also contained four jars of a strange reddish powder.

The older friar had to restrain Seiler at this point because he was so angry and disappointed that he wanted to throw the whole chest away. But the friar suggested that they should closely analyse the parchment and try to decipher whether the parchment gave any clues as to the significance of the strange powder. The aged friar had learned more, and read more about alchemical practices, than Seiler and came to the conclusion that the red powder could be the transmuting compound that he had heard of. Conspiring with Seiler, the friar persuaded him to take an old tin plate from the abbey's kitchen and they then covered it with the reddish powder from one of the jars.

They then heated the plate in the fire and as if by some magical process, the tin plate shortly became solid gold. It is hard to imagine any scientist that could possibly explain how a powder of any kind could change the atomic number of tin to that of gold, simply by heating it in a fire, but we could refer again to the Count St Germain who in front of a witness covered a coin in a dark powder and when he heated the coin in the fire it became pure gold. The Count St Germain was a proven alchemist. However, Wenzel Seiler was so happy with the result of the experiment that he could not wait to get into town to sell the, now gold, plate which he eventually sold for twenty ducats. The old friar was a bit concerned however, as there was a definite risk of an informer enlightening the Abbot that a member of the order was seen selling the friary's property.

The story continues. Not long after this incident, the old friar passed away to leave the young Wenzel Seiler the sole owner of the alchemical data and the amazing powder. However, in order to increase his wealth, he had to first find someone in whom he could confide regarding his discovery, and secondly to find a way to leave the monastery.

Eventually, he confided in another young monk called Francis Preyhausen. Preyhausen was amazed to learn about the powder and its capability, and so the two monks made plans to depart the abbey in the coming spring. Before this however, Wenzel used some of his ducats to purchase wine from the town and enjoy a visit from a young cousin of his from Vienna named Anastasia. The abbey was evidently quite loosely run if the inhabitants were allowed trips into town and company. And not surprisingly, information regarding the stolen plate, and the ducats, and the buying of wine, soon reached the ears of the sleepy Abbot who eventually summoned young Seiler in for questioning.

The result was that the Abbot took a group of friars and Seiler to inspect Seiler's cell which they found strewn with empty wine bottles and Seiler's cousin, Anastasia, naked on his bed. During a few embarrassing moments during which time the girl managed to wrap herself in her cape, the elders of the friary proceeded to deliver to her a sermon regarding the dangers facing her soul. For his punishment, the young Seiler was flogged and bolted in his cell, which only made him even more determined to escape. His friend, Francis Preyhausen, whispered to him through his cell window to hand the over the jars of powder through the bars – at which point, Seiler must have been rather concerned that Preyhausen may bolt with the powder. Then, to his further dismay, Seiler was transferred to a prison cell and the future began to look very dark.

However, it turned out that Preyhausen did not desert Seiler and was not idle in formulating plans for their escape – which he achieved. This was followed by an adventurous journey during which the young monks soon realised how dangerous their lives could become with this secret and the potential that they held in their hands. It became obvious that Francis Preyhausen was the more intelligent of the two and took precautions to protect the powder and to find the right connections. Ideally, they would find someone of influence who also had an interest in, or even better was a practitioner of, alchemy. Eventually they secured the patronage of a certain Count Peter von Paar who was an ardent student of alchemy and a friend of Emperor Leopold I of Germany, Hungary and Bohemia (1640–1705). An audience was arranged for the two young monks with the emperor who was just as interested in alchemy as Von Paar and the two monks.

During this audience, and in the presence of a Father Spiez and a Dr Joachim Becher, Wenzel Seiler transmuted a certain amount of tin into pure gold over a period of about quarter of an hour. A written declaration to the effect was signed by all the witnesses. However, after witnessing this miraculous transformation, it became clear that Count von Paar was

Make Me some Gold

a slippery character and betrayed his friendship with Seiler and Preyhausen by robbing Wenzel of the transmuting powder at gunpoint.

Somehow, Seiler recovered the powder, and von Paar met his death soon after (it is not recorded how). Another transmutation soon followed after which Seiler gained the patronage of Emperor Leopold I and also of Count von Waldstein who was captain of the bodyguard. Eventually, the Emperor himself manufactured alchemical gold with the use of Seiler's powder. In 1675, it is said, a special ducat was forged that was imprinted with the image of Leopold I and was made from the gold that the sovereign had himself alchemically produced. On the reverse side of the ducat there was the following inscription: "with Wenzel Seiler's powder was I produced and transformed from tin to gold."

Seiler's abilities and reputation were gaining ground, and further successful experiments in alchemy were conducted by Seiler at the Palace of the Knights of St John in the Kärtnenstrasse in Vienna. A gold chain was made from this alchemical gold on the orders of Count von Waldstein, and on 16 September 1676, the Emperor Leopold knighted Seiler as "Von Rheiburga" (Rheiburga being the maiden name of Seiler's aristocratic mother) and he was then officially appointed as the "Court Chemist" to Leopold I. At this point however, the red powder that had served him so well, was almost gone. Both Wenzel Seiler and the Emperor Leopold himself made strenuous efforts to multiply the powder but without success.

We now remember the gold medallion in the Vienna Museum – the lower part of which is pure gold. In 1677, there was sufficient red powder left for Seiler to dip this large medallion into the transmuting compound; and as a result, its lower two-thirds turned it into pure gold and the notches that are now evident in its side were created when its purity was tested. This test was made on the request of Professor A Bauer of Vienna in 1883 and it was found to be pure, solid gold. The fact that this case of alchemical transmutation is recorded in relatively recent history, must offer strong evidence for the existence of this practice in former times.

However, the greatest mystery is in the transmutation itself; while a great effort was made to ascertain the purity of the gold, no attempts seem to have been made to surmise what ingredients, compounds, or minerals the powder might have consisted of. This seems strange and only adds to the mystery. There is a nineteenth-century painting by the Polish artist Jan Matejko which dramatically portrays a supposed alchemical transmutation that was carried out by Michael Sendivogius in Kraków, Poland, and performed in the presence of King Sigismund III of Poland early in the early seventeenth century.

Alchemy was not confined to just producing gold; some alchemists also claimed that they could produce gems. I have covered this alleged capability with regard to the incredible Count St Germain in *Strange Realities*; he was apparently able not only to produce diamonds of the finest quality but also to remove any flaws thereby increasing their value. He performed this latter feat on a large diamond that was owned by King Louis XV, and that the king's jeweller confirmed was the Kings property.

Modern science has produced diamonds using a number of different methods. It has transformed ashes or anthracite into diamonds using heat and pressure – a process that has been used to create diamonds intended for industrial purposes. In 1969, Dr Willard Libby, a Nobel Prize winner, demonstrated a method whereby he sandwiched a block of graphite between two nuclear devices. But to produce and cut a flawless diamond for jewellery is an extremely expensive process – so one might as well buy the diamond.

But traditional alchemists were also in existence relatively recently. For example, in 1897 Dr Stephen H Emmens, a British physician living in New York, claimed that he had discovered a method for transmuting silver into gold. He was as good as his claim and between the months of April 1897 and August 1898, he produced (and sold) more than $10,000 worth of gold to the US Assay Office, situated in Wall Street. At the time, the New York Herald printed the following headlines: "This man makes genuine gold and sells it to the US Mint." But, although the Assay Office did admit to buying the gold, the newspaper could not resist indicating its doubts by adding "Did this man actually manufacture gold out of silver as claimed?"

Perhaps the question of whether or not transmutation could be accomplished is of lesser importance than the question of how it was that the ancient alchemists could discover these things in the first place. How could these alchemists have envisaged the chemical process that transforms one element into another? How did they know to use the metals that have such close atomic numbers – mercury, lead, gold etc – without allegedly having any knowledge of the periodic table? Even if the alchemists did not succeed in making gold, they still seemed to anticipate the concepts of future science with regard to the essence of matter.

Even science itself excluded the possibility of transmutation until Curie, Rutherford and other visionaries came along. And after this, it is noteworthy that the Nobel-Prize-winning chemist, Dr Frederick Soddy – who came up with the theory of isotopes, wrote *The Interpretation of Radium* (in 1909) and pioneered nuclear physics – did not (although considering it) deride alchemy.

Nevertheless, nothing has been proven, and it must still remain a "noisy bee" in the bonnet of science, when attempting a serious investigation into alchemy, to hear that gold can be produced by merely dipping silver into a powdery substance and putting it into a fire. But this is exactly what seems to be required, and is a process that is said to have been performed, in front of observers, many times over the centuries. We must, of course, bear in mind that magicians today are capable of wondrous things that are totally unexplainable, even up close. Yet, magic could not have filled royal coffers with gold, or built and renovated churches throughout France.

Egyptian mythology pointed to Thoth, Hermes or Mercury as the culture-bearer who had revealed to mankind the hermetic arts – one of which was alchemy. Hermes (or Mercury) was also the founder of medicine. As Andrew Tomas put it, "It is on the rock of hermetic science, that modern medicine is built. It is fascinating to trace the stream of medical science from pre-historical medicine man, herbalists, magician, priest, to the pharmacist and the doctor of modern life." But to whom do we trace the art of gold-making? To gods? Perhaps to gods who are "not of this earth"?

CHAPTER VII

ODDITIES OF EARTH AND COSMOS

In the occupations of astrophysics and of astronomy (which in certain respects are closely linked) experts are often kept guessing. Ideas and theories that are accepted as established and observable fact, have often had to be altered, modified or disposed of altogether. Since the period of the earliest telescopes, the universe, or more precisely our assumed knowledge of it, has been in a state of flux. And every so often it appears as though some hidden power suddenly states "Hold it, you had better start again, everything you thought you knew about the universe is wrong." Many theories and textbooks have had to be changed during the long human interest in things celestial. It has been stated that the universe is not only stranger than we imagine, it is stranger than we can imagine; and we can almost hear the astrophysicists saying, "I'll drink to that".

Although the earliest astronomers were actually the ancient Babylonians, Egyptians, Incas and the Arabs, we usually think of much later figures such as Galileo Galilei and Giordano Bruno when we think of the pioneers of this field. And these figures bravely discussed their theories in the face of authorities who did not always want to hear them. In fact, Giordano Bruno was executed for his theories and Galileo only narrowly escaped this fate. Thankfully at the very least, the modern counterparts of Bruno and Galileo do not have to "watch their backs" while being bombarded with all the puzzles, mysteries and oddities that crop up during their observations.

An astronomer name Asaph Hall, who was the director of the Naval Observatory in Washington in 1877, is said to have been the first person to discover the two moons of Mars. These two moons are known as Phobos and Deimos. However, there seem to have been many earlier allusions to these moons. For example, the fifteenth book of the *Iliad* alludes to the fact that Mars had two companions that were known as Phobos and Deimos; and in ancient tradition they were the Martian satellites expressed in symbolic form. And nearly 300 years before Hall, between 1571 and 1630, the famous astronomer Kepler left the following solution to an anagram attributed to Galileo, "Greetings to you, the twin offspring of Mars." So, it is almost as though the renowned early astronomers knew or suspected the existence of the Martian satellites but preferred to keep their discoveries quiet.

Moreover, Cyrano de Bergerac (1519–1655) also mentioned the two moons of Mars in his *L'Autre Monde*. And Cyrano was not alone, even Voltaire, who many of us view as a philosophical writer rather than an astronomer, was also quite certain that Mars possessed two satellites. In his 1752 *Le Micromégas* he wrote, "Coasting along the planet Mars, which is well known to be much smaller than our own world. There are two moons that are subservient to that orb, which have escaped the observations of all the astronomers."

Even Jonathan Swift made astronomical references in his novels; for example in *Gulliver's Travels* he described the flying island of Laputa that was propelled in space by a magnet. And in his story, a group of scientists on this weightless space platform speak about the two moons of Mars. One of these "lesser stars or satellites" as Swift called them, orbits Mars at a distance of three martian diameters from the planet's centre and the other whirls around it at a distance of five diameters. While the dimensions, sizes and distances that Swift used differ from the actual ones, it is notable that he was writing astronomical knowledge that was only "discovered" by Asaph Hall nearly 150 years later. Dr I M Levitt also mentioned that, "Although the data mentioned by Swift regarding the dimensions and distances etc., regarding the satellites was not strictly accurate … this similarity between the hypothetical satellites and the actual ones was so close that it remains one of the most amazing feats of speculation."

Perhaps Jonathan Swift obtained his information by studying ancient documents, yet, how were these documents written if, as it is assumed, the ancients did not possess telescopes? In fact, it is likely that this assumption is false because ancient depictions of astronomers looking through telescopes exists. And it is established that many observatories existed in ancient times – so why build observatories if there was nothing to observe with?

One of the Moons of Mars, Phobos, is the fastest travelling Moon in the solar system; it whirls around Mars in seven hours and thirty-nine minutes – which is faster than Mars itself revolves on its axis. It is stated that this phenomenon is without parallel in our solar system.

The planet Venus also has mystery attached to it – the idea of the moon of Venus. To date, our modern astrologers are not aware of any moon orbiting Venus. And no astronomer of the 1600s ever envisaged the planet Venus as possessing a moon. Yet, early in the morning of 25 January 1672, the renowned astronomer Giovanni Domenico Cassini (who was aware, for example, of the Rings of Jupiter as well as its great red spot) sighted a small object near Venus that shouldn't have been there.

Cassini did not get too excited straight away as he wanted to allow for possible error; he knew that he had discovered something unusual but did not wish to cause a sensation by immediately staking a claim on the discovery of a Venusian moon. He saw this object again fourteen years later at 4.15am on the morning of 18 August 1686. We can only assume that in the interim he looked for it in vain and had to wait until this early morning in 1686 before he saw it again. Cassini was probably glad that he had not announced his original discovery as it had somehow disappeared. And in fact, it seems that the second sighting was not the same object at all because it proved to be much larger than the object originally seen in 1672. This second body was estimated to be about a quarter of the size of Venus and was situated at an approximate distance of around three fifths of the planet's diameter. The detail of this sighting seems to indicate that Cassini was using quite sophisticated, or we could say "top of the range" optical equipment for 1686. It was observed that this new Venusian moon went through "phases" during the month from crescent to full, as of course does Venus itself as well as our own Moon.

Cassini studied this strange object closely and naturally recorded the data meticulously in order to assist other astronomers whose corroboration would be needed in order to confirm his discovery. Naturally, other astronomers then also renewed their interest in Venus and began to notice other strange occurrences. For example, on 25 October 1740, James Short of England discovered another astronomical body near to the planet Venus. Short's object was estimated to be one third of the diameter of the planet. He continued to observe this new "visitor" through his telescope for around an hour. Then, on 25 May 1759, yet another astronomical body was sighted by Andreas Mayer of Greifswald in Germany. He observed it for about half an hour while it was observable in the vicinity of Venus. But it seems strange that in the decades between Cassini's sightings and those of Short and Mayer, that no regular orbit of any moon could be established. And all the while, the newspapers and scientific journals were likely to be reporting on these strange events and speculating on why all these periodic visitors might be appearing but not staying.

So naturally, during the interim years, other astronomers would have become quite sceptical about it all. But three years after the observation of Andreas Mayer, in 1761, Jacques Montaigne, a member of the Limoges Society who had been sceptical of all the Venusian Moon observations saw it himself. He identified it as a comet and observed it in that March on four separate occasions. And then it emerged that on February 10, 11 and 12 of that same year, Joseph Louis Lagrange of Marseilles (who later became the Director of Berlin Academy of Sciences) had also reported observations of a Venusian "visitor".

Lagrange also saw it on repeated occasions – again on March 5, 26, and 29 of that same year. And then, to add to the lengthy list of observers that year, a man named Montbaron, of Auxerre in France, also spotted an object orbiting Venusian through his telescope. Then, an astronomer named Roedkider, in Copenhagen, made eight observations during the months of June, July and August of that same year.

Finally, all these observations came to be recognised by the authorities and Frederick the Great, King of Prussia, proposed that this new moon of Venus be named "d'Alembert" in honour of the French Savant Jean-Baptiste le Rond d'Alambert. It can be assumed that the King of Prussia would not have made such a suggestion unless he had been shown it himself by an experienced astronomer; he would not have wanted to appear foolish for suggesting the official recognition of a "moon", only for it to disappear. Yet, as it turned out, this is precisely what happened. Although it was sighted again seven years later on 3 January 1768 by Christian Horrebow in Copenhagan, it was not seen again after this until 1786!

So how could a "moon" disappear for eighteen years? If this happened today, ufologists would certainly come up with an explanation – likely along the lines of it being a visiting cosmic craft from another world paying periodic visits to carry out surveys of Venus and perhaps of Earth also. To be sure, when contemplating the strange behaviour of this appearing and disappearing object, this explanation seems as good as anything else on offer as some "force" must be controlling any object's disappearance and reappearance.

For all our sophisticated technology, we still don't know what these strange bodies are that appear to be freely moving about in our solar system. They could certainly, in some cases, simply be asteroids. Yet, asteroids hold a steady course until they hit something such as a planet or another asteroid; they do not appear and reappear in different sizes. The normal celestial mechanics of planetary and moon formation appears to ensure that they are round globes, but the moons of Mars are far from that description and are almost certainly captured asteroids – possibly from a world that may once have existed in the vicinity.

In the case of the Venusian mystery, the jury is still out. Certainly, any feasible theory that conforms with logic and things celestial, is worth consideration. And even theories put forward by ufologists should not be simply written off as "new age" thinking – in part at least because these ideas have also been proposed by science. These theories suggest that the objects are hollow, contain beings from elsewhere, and that they travel through the cosmos as self-sustaining vehicles, tapping into solar energy. Perhaps they study other worlds and land on planets sometimes, all the while increasing their vast data banks of knowledge.

It has even been proposed that our own Moon arrived in earthly orbit and is in fact (or was) a purposely steered and occupied object. Two scientists from the Soviet Academy of Sciences actually proposed this theory. I have covered this idea in my book *When the Moon Came*. However, we return to the problem of whether or not there is a moon orbiting Venus. Perhaps the most obvious explanation is that the objects sighted in the seventeenth century were asteroids that simply disappeared when they collided with another planet or when they disappeared into the Sun. If an asteroid originated in an asteroid belt, then it would have returned there if it was not stopped by another cosmic body. Or, if it collided with another object in the asteroid belt, then it could have been set on a new course – which of course might be a worry for our own planet. So asteroids being thrown off-course might account for one-off appearance, but not a reappearance; any celestial body that reappeared (if proved to be an earlier visitor) after a long absence is certainly a puzzler.

It is comparatively recent in cosmic timescales that the existence of other worlds has been under consideration. But these thoughts must also have occurred in ancient times. Even before the teachings of Jesus, the ancient Greeks were making very sophisticated statements in relation to this idea, and educated Romans must have been aware of the writings of these predecessors. When Pontius Pilate was interrogating Jesus, and Jesus stated that his Kingdom was "not of this world", why was it that Pontius Pilate was not immediately intrigued (a question I also asked in *Pillars of Fire*)? Why did he not ask Jesus which world was he from? Or how far away is it? He could have learned much, but only if Jesus had cooperated. One supposes that he would either remain silent or state the biblical equivalent of "that information is classified."

It has been said that if there were only barren and desolate worlds circulating all those trillions of stars, then it would be an awful waste of real estate. If our universe is the result of a "creation", then it would surely would have been a waste of time for any divine creator to make only such a tiny part of it living. Even if it was a big bang event – one small living planet seems an unlikely coincidence. Although we now know that there are huge amounts of other worlds orbiting other suns, we wonder what the point of their existence is if they are completely barren and inhospitable to life? Perhaps the answer lies in the fact that they contain resources to be plundered by beings such as ourselves. After all, the Moon is desolate and barren but that world will likely be the first one that we will pillage and plunder for its resources (if other beings have not already done so). It would be quite disappointing for astrophysicists to discover another planet that orbits a sun-like star, situated in their ecologically-ideal "goldilocks zone", possibly containing water, only to find that it does not fulfil all the pre-requisites for life – perhaps being too large for example. Yet this disappointment seems often to occur in the case of newly-discovered planets.

However, with the search on-going, the odds are in favour of discovering an earth-like world. Perhaps even then our hopes will be dashed again when we contemplate the immense distance we would have to travel to reach it – but those problems will no doubt be overcome when science fiction once again becomes scientific fact. If this had not happened in the past, then we would not be in the position we are now in, that is, on the springboard of the next giant leap for mankind which may happen sooner that we expect. With the amazing advancements that are being achieved in the area of artificial intelligence – far surpassing the rate of advancement in other scientific areas – then it may not be "mankind" that makes this giant leap, but mankind's representatives in the form of entities that are part-biological and part-machine.

But again, we return to Venus. This time to an astronomer named Jean-Charles Houzeau (1820–1888) who seven times observed an object in the vicinity of Venus. In 1884 he baptized the object "Neith" in honour of the Egyptian goddess of learning. So we ponder again upon the question of where "Neith" had been hiding for a century.

On 3 August 1892, the American astronomer Edward Emerson Barnard also sighted a seventh magnitude object in the vicinity of Venus. Barnard was at the forefront in the search for other planets at the time and had discovered another large planet near a sun that was named "Barnard's Star" in honour of him. So he was clearly an eminent astronomer, and in spite of the fact that he had no faith in the previous claims of a Venusian Moon he clearly thought that the idea was at least worth a look. His report on his sighting of the Venusian object commanded some respect and was considered highly reliable because of his previous accolades – such as the star in the constellation of Ophiuchus mentioned above and his discovery of the fifth moon of Jupiter.

Yet, this mysterious moon of Venus once again disappeared. Evidently astronomers were not deterred and carried out a sort of vigil for another one hundred years but without success. Astronomers often learn of unexpected incidents simply by observing whether there are any perturbations from the usual orbits – and then exploring the possible reasons for them.

Another significant cosmic discovery occurred in the 1930s with the discovery of the planet Pluto. This occurred when the mathematician, Dr Clyde Tombaugh, figured out that the unusual movements of the planet Neptune, must mean that there was a planet beyond it. And sure enough, when he pointed his telescope in the right direction, his calculations proved correct and he found Pluto. Yet there is an oddity here. Now, after all these years of further astronomical research, the current thinking states that the planet Pluto could not have been of sufficient mass to have any detectable effect on the orbits of Neptune or Uranus. And so the assumption on Tombaugh's discovery is that it must have been pure coincidence and nothing else. Perhaps this teaches us that sometimes it is useful to make errors in our calculations (if indeed errors were made). Either way, it seems that the cosmos does not reveal its mysteries in a consistent way, or only very carefully.

This is born out in another, earlier, oddity that occurred in the nineteenth century but is worth a mention here as it was considered a major scandal in celestial circles at the time. On 20 March 1859, a certain Dr Edmond Modeste Lescarbault, of Orgères-en-Beauce, France, observed a moving body, that appeared to be crossing the disc of the Sun (not unlike the way in which the Venus event occurred). Lescarbault tracked the course of this object for a period of one and a half hours which rules out that there might be any natural explanation, such as an insect for example. One might suggest that it was a "sunspot" – a dark patch, or patches, that appear periodically on the surface of the sun. An astronomer by the name of Urbain Le Verrier was the Director of the Paris Observatory at the time and he was sceptical of Lescarbault's observation – possibly because he thought that Lescarbault was only interested in enhancing his reputation by making claim to the discovery of a new planet.

So Le Verrier arrived at Lescarbault's observatory with little enthusiasm and plenty of scepticism. We might ask ourselves why, if Le Verrier had access to the likely more powerful telescopes at the Paris Observatory, did he not immediately check Lescarbault's observation himself rather than travel to Orgères-en-Beauce. Perhaps professional rivalry meant that he was reluctant to acknowledge Lescarbault's discovery (such as occurs, for example, in the world of paleoanthropology where researchers would dearly like to have their own name linked to a fossil to prove human origins). But Le Verrier, after checking out Lescarbault's observations, could not deny any further that something of importance had been observed, and concluded that and "inter-mercurial" planet had been discovered. So Lescarbault gained the satisfaction that he had "made his mark" so to speak.

Between them, the two astronomers computed its mass to be a fraction of that of Mercury – a tiny world that would surely be better described as an asteroid than a planet. They figured that it must be orbiting the Sun at a rate of around once every nineteen (of our) days, and they named it "Vulcan". Buoyed up by this discovery, Dr Lescarbault did his calculations and presented them to the Academy OF Paris in January 1860.

Most countries like someone to emerge in their land that they can honour internationally, particularly in the world of scientific discovery, and indeed, Napoleon III was delighted to award Lescarbault the highly coveted award of "Légion d'Honneur". However, while Napoleon and France where basking in the glory of this astronomical discovery, "Vulcan" (its mission perhaps accomplished) promptly disappeared. After a frantic search through astronomers' telescopes everywhere, it was not seen again – it completely vanished just as unexpectedly as the elusive "Moon of Venus".

So a "scandal", in the form of professional embarrassment, occurred, and it was felt that the authorities were too quick to claim the kudos of a discovery. Yet, the mystery of what exactly was observed, on all of these occasions, remains. When considering all of these mysterious events in total, it seems to point to the idea that a fleet of celestial craft from another world are (or were) cruising around our solar system analysing our star and its attendant bodies at their leisure – some large missions and craft, some small, depending on their mission profile.

To add to the Vulcan affair, and to complicate things even further, in 1878 an astronomer from the University of Michigan, USA, named Professor James Watson, claimed to have seen not one but two Vulcans. It is not clear how efficient the optical equipment was during the time of these observations, but it was suggested that it may have been a trick of the light – a reflection – that caused this double Vulcan sighting. This seems likely because another astronomer also observed Vulcan but, like Lescarbault, also only saw a single object. This astronomer's name was

Lewis Swift, and he was able to get a good observation of what he thought was the same object as Lescarbault's. Swift's observation was made from the Pikes Peak Observatory in Colorado. Although Lewis Swift was described as an amateur astronomer, he was no ordinary star gazer as his work on Nebulae had previously received widespread recognition by those in the astronomical sciences. The disappearance of Vulcan must of course have been something of an embarrassment for Lescarbault, but there is no evidence that Napoleon III asked for the award back.

The behaviour of these travelling bodies, whatever they are they, do not fit the description of asteroids because of their odd habit of appearing and disappearing, and so the extra-terrestrial assumption may be as good as any other at this point. Something appears to have been cruising around our solar system during those times, and perhaps it is now "mission accomplished". And this idea of the existence of other worlds is not new – it goes back not just for centuries but for millennia.

Eventually, we achieved the technical ability to closely observe for ourselves the noticeable dimming of a star's light, and realise that it indicated that something was passing in front of it – probably a planet. And although mathematical calculations could ascertain that the perturbations of an object were caused by the effect of some other mass – now, we can gain definitive proof of such things with the use of imagery taken from orbiting telescopes.

An intelligent observer on another world looking up at its night sky may see our Sun twinkling away just like millions of others but there would be nothing special about it. Such an observer would have no idea that the third planet from that star would be teeming with life just – just as we also have no idea of the worlds that might exist in the orbits of the stars in our night sky.

And so in the vastness of the universe, it would seem quite incredible if there were not other worlds out there. To discover life in the form of intelligent beings with pointy ears and American accents would indeed be a fantastic event, but we have reached the point that if we discovered life of any type out there, we would be amazed and delighted. This would fulfil the expectations held by humanity since we first looked up at the stars. Our current advancements in astronomy – from the Hubble Telescope to the more sophisticated planetary search satellites – have, and still are, discovering other worlds that can be described as earth-like. As previously mentioned, we sometimes read of exciting claims that a planet like Earth has been discovered, and our minds make great leaps forward, imagining blue skies, oceans, and intelligent beings. But our hopes are often dashed when we receive further information on the real conditions existing there. After our imaginations have conjured up earthly scenes of green grass, singing birds, and other possible life forms, we are soon brought to reality by getting hard facts. We learn, for example, that this new "earth-like" world is twice the mass of our own so that even though its star is similar to ours, and it orbits in the favourable "goldilocks" zone, any beings living there would be huge and massively built in order to cope with the extra gravity etc.

But we might question the reality of such pessimism. It is true that our own astronomers bound about on the Moon so freely due to the lesser gravity – and so we might deduce that visiting a world twice our size means that we would hardly be able to take a step forward. Yet, our own planet has given birth to great lumbering creatures that dwarfed humans in stature and bodily mass and that existed for many millions of years. Bodily mass no doubt plays a role. Water is quite heavy and we are composed of two thirds of it in our makeup, but what if it this water was more diffuse and less densely packed? Then perhaps we need not be so huge for a bigger planet after all.

And what about the role of evolution? If we assume that life should exist on other favourable planets, then it becomes even more likely when we consider that life on our own planet has evolved from beings such as the dinosaurs, to animals, to intelligent beings who are able to contemplate travel to the stars – all in a relatively short period of time. If all that is needed is time, and when we consider that even just the existence of the dinosaurs is estimated to have lasted for one hundred and eighty million years, then we might certainly allow for a whole host of intelligent beings to have evolved on other planets.

And although the size-to-weight ratio on planets that might occupy the right orbital position but are far larger than our own, might be different, we should remember that on our own planet, the huge dinosaurs lived alongside tiny creatures such as insects. And of course, the oceans at that time were also replete with marine life ever since the rather miraculous event that we call the "Cambrian Explosion" occurred.

It is unlikely that the cosmic mysteries will resolved in our lifetimes, but the likelihood of this occurring increases as we begin to venture into the universe ourselves. Due to the obvious curiosity embedded in the human psyche, we can congratulate ourselves a little when we consider the mass of information that we have gained, even before we left the planet at all. And yet, when we reached the Moon, a completely new set of mysteries and problems confronted us. We might wonder at how contented and relaxed we would be if we were just like our alleged primate cousins – never looking up at the sky and even if we did, soon losing any interest (due to our short attention spans) and turning back to poke at an ant's next for desert. For humans, the wonders of the universe – the question of other life-forms existing out there in the mass of stars, planets and moons – has existed as a form of scientific thought ever since our most ancient of ancestors.

One of the biggest problems that will confront us when we set ourselves upon the Moon or Mars will be radiation and the threat of cosmic debris. This is something that we on Earth, with our protective atmosphere, rarely have to worry about. It is estimated that about three major impacts per year occur on Mars – no doubt mainly from the asteroid belt that were it not for our protective atmosphere would be our "neighbour from hell". And as mentioned, radiation is another problem that no doubt astrophysicists are already working on.

As long ago as 3,500 B.C. Egyptian star charts were in existence. This indicates that they already had a systematic study of astronomy and, amazingly for their time, it is clear that they were aware of Mercury and Venus being closer to the Sun than the Earth, Mars, Jupiter and other planets. Obviously, the possession of star charts in the first place indicates in-depth study of the observable universe. In their turn, the Babylonian priests of 2000 B.C. were recording their observations of the orbits of Venus, Mars and Jupiter in their cuneiform method of writing. An indication of the advanced knowledge of astronomy held by the Mesopotamians, is that they were even more advanced with regard to the forecasting of eclipses than the Egyptians who were ahead of them in other branches of the astronomical sciences. The Babylonian priests were able to forecast the eclipses quite accurately.

And the question remains, did these ancient practitioners use telescopes? Some scholars of the ancients believe that they simply could not have advanced so far in their astronomical knowledge without them – particularly when viewing the moons of distant worlds. And yet, it is difficult to imagine the ancients being able to produce lenses for telescopes that were accurately ground. We could apply this to all "ooparts" ("out-of-place-artifacts") – to any intricately manufactured and designed objects that appear to have originated a long time in the past.

Such objects ought to be the subject of a prolonged investigation in their own right, but the fact is, that they are something of an embarrassment to science. There is a myth that is still not completely extinguished, although it ought to be, that the Romans, out of the kindness of their hearts, civilised Britain and prior to that we were all untamed and wild savages, daubed with woad. The fact, on the other hand, is that the ancient inhabitants of England were just as proficient in astronomy as the ancient priests of Egypt or Sumer. The computations evident in the structure and alignment of Stonehenge – the Stone Age structure in Wiltshire – have revealed a very precise knowledge of the solstices and equinoxes, and the ability to accurately predict eclipses. The ancient builders of the megalithic structures clearly had this knowledge when the structures were built – around 4,000 years ago.

To gradually amass all this knowledge into a working technology was not bestowed upon the ancient Britons overnight – it must have taken enormous time periods for it to have evolved into such an exact art. Moreover, precise measuring instruments and a knowledge of mathematics must have gone hand-in-hand with such accurately aligned structures. The question is, was this development indigenous or was it the result of instruction of some sort given by ancient tutors from knowledge originating with the antediluvians – that is, originating from before the flood. Was this knowledge spread to the people of Britain by those post-diluvian people who, as the Bible would have us believe, emerged from the Ark of Noah to reinvigorate the Earth and its technology after the flood?

Celestial knowledge, along with many other subjects, was prevalent and well-established in ancient South America, Egypt and Babylonia, and was written down into encyclopaedias by the ancient Greeks. The Greeks enhanced this knowledge, making it into a fine art. Their extremely knowledgeable and educated philosophers existed in very large numbers and most of their names are still renowned today. In fact genius per capita seems to have been far more widespread in ancient times than it is today, but then, we must perhaps consider the enormous population increase since those times. The imaginative thinking of the ancients was on par with modern-day planetary scientists and astrophysicists. Their conclusions were in many cases accurate and, unlike as occurred in the Middle Ages, they were not restrained and stultified by ecclesiastical fears of oppression. There was no fear in classical times of death for merely proposing a philosophical idea that might divert the people away from the approved rules of the church – and of course, this oppression was the whole cause of the ignorance that prevailed in the Middle Ages.

However, these rules were brought about by men and their enclosed thinking – and so we might imagine that "God" would not have wanted this, nor his son Jesus who, after all, stated that he was "not of this world." We might argue that Jesus was born of this world entirely due to the actions of others who were also "not of this Earth." We might argue that the birth of Jesus occurred as a result of the artificial insemination of his mother by those beings, and that this is the reason that he was named by them "the son of man." Jesus, if the stories in the Bible about him are true, would surely not have condoned the murder of all those helpless women for the enjoyment of psychopaths such as Mathew Hopkins, the "witch finder general", or the monstrous actions of the inquisition. It was not "gods" but humans dressed in holy robes who carried out these actions that they must have known were entirely against the teachings that they were blatantly disregarding – such as "Love they neighbour", "turn the other cheek", "judge not lest ye be judged."

How much further advanced would technology be today if these restraining influences had not occurred? The fact that the Earth is round and revolves around the Sun was already proclaimed by the Greek Anaximander in 500 B.C. Most certainly, if he had made such a statement in the Middle Ages during the time of the Inquisition then he would have been consigned to the flames. The equally brilliant Pythagoras confirmed that "the Earth is a globe." And in 200 B.C. Aristarchus of Samos stated that the "Earth travels in an orbit around the Sun and at the same time rotates on its axis," and added, "as do all the planets." How accurate is this for its time? And in the same period, around 200 B.C., Hercules of Pontus knew and stated that the Earth turns on its axis once every 24 hours.

What an appalling loss, or suppression, of knowledge must have occurred over the following centuries. What if it had been allowed to continue unimpeded? We may by now have travelled far beyond our solar system. Perhaps we would now be preaching our knowledge to lesser mortals on a planet far away, which for all we know is how we indirectly received it.

Oddities of Earth and Cosmos

The ancients of other cultures seem to have been equally advanced, well-informed, and unimpeded by exterior forces. The sacred book of Hindu culture, the *Rig Veda*, contains a passage that refers to "the three earths" – one within the other. And we now know that it is indeed a fact that the Earth has three zones – that is, the inner core, the mantle (or outer core) and then the outer crust.

Beginning with the maritime explorers of the fifteenth century, and with the renaissance of ancient knowledge and the use of extremely ancient maps that had been copied and re-copied since the time of Alexander the Great – knowledge began to improve again. Christopher Columbus, whose name has been preserved through history as a great explorer, preserved many ancient charts. During his trips at sea on dark crystal-clear nights, he observed his share of celestial oddities – he even observed what we would define as UFOs. On one occasion, Columbus and one of his sailors, observed a light that was obviously under some kind of control. It "buzzed" his ship, "passing by in sudden gleams."

Columbus also made what have turned out to be very accurate observations about the earth; in a letter that is preserved in the city of Madrid, he made a statement that concerned the shape of the Earth – stating that it was slightly "pear-shaped". This is a fact that no one else at the time was aware of, and that has since been confirmed as accurate by our modern-day orbital geographical satellites. He must have gleaned this knowledge from all the ancient maps, charts and data that he had amassed and studied. Apparently, he had quite a collection of them, some of which were apparently so old, and went so far back into antiquity, that no one could explain their origins.

Another area of understanding in which people in history seem to have had advanced knowledge of, was the Moon. There is an oddity concerning our Moon that our modern-day astronomers are aware of, and that is known as the "lunar wobble". But already in the tenth century, an Arab astronomer called Abu'l-Wafa wrote about the variation of the Moon and its "wobble". This phenomenon cannot be detected without the use of special instruments – such as a chronometer – that we usually assume the astronomers in history didn't possess. The discoverer of this lunar variation is usually attributed to the astronomer Tycho Brahe (1546–1601) – but if Abu'l-Wafa also discovered it and it is also highly likely that it was known about in ancient civilizations, then the jury is still out.

And when it comes to the gravitational effects of the Moon, we know today that the tides are an effect of the Moon, and we now know that the ancient sages of China also knew this. But when the German astronomer Johannes Kepler stated this around the end of the sixteenth century, he was severely censured. Like many other scientists and experimenters of his day who were in danger of persecution for resurrecting the knowledge of antiquity, Kepler had to tread carefully. This kind of ignorance with regard to the Sun, the planets and astronomical matters in general, long prevailed in the late Middle Ages.

The ancient Maya had a more precise figure for the length of the year than our own Gregorian calendar. The Mayan calculation for the amount of days in the year is 365.2420 days, and their figure for the length of the month is 29.53086 days. So again, we must ask, how did they reach such precise figures without the instruments necessary to compute these figures? The ancient astronomers spoke-of and recorded many facts that they could not possibly have known about without the use of telescopes.

The telescope was thought to have been invented in 1608, so when Galileo "discovered" the moons of Jupiter in 1610 he did so with a telescope. We therefore ask, how did the ancient priests of Babylon know about the moons of Jupiter when they cannot be seen without a telescope? Professor George Rawlinson argues that there is strong evidence that the Babylonian astronomers were also acquainted with the seven satellites of Saturn. Yet in our own recorded history, these "satellites" were supposedly only first discovered by Cassini, Huygens and Herschel, from the time period around 1655 and later.

Oddities of Earth and Cosmos

It has been speculated in the past and also fairly recently, that the history of humankind may be far more ancient than we currently imagine. And just as in the case of theories in astronomy, the history of mankind has had to be continually altered and revised, or occasionally disposed of all together. However, historians seem even more reluctant than the astronomers to accept this and often cling tenaciously to old ideas and theories, as any thought of having to re-write great parts of history is most unpalatable to them.

Many theories have been put forward or speculated upon to account for the disturbing implications that arise when artifacts are found that seem to indicate the ancientness of human kind. At the end of this chapter, we will expand on this. It has even been speculated, with the assistance of some accounts in ancient folklore, that humankind arrived on the Earth as extra-terrestrial migrants in far ancient times and that our current technological advancements are a subconscious attempt to revive what we once knew – and to revive that knowledge that was brought to earth by those beings, who are of course, our ancestors. Of course, there is nothing new about these ideas.

Some accounts have speculated further that our life forms may once have existed on planet Venus, and that when we had ecologically despoiled Venus to the point that it was no longer habitable, we (avoiding Earth and all the predators) settled on Mars. Humans seem to possess a most undesirable trait of destruction – and either end up destroying each other in war, or destroying the planetary environment. And we may have possessed this feature from time immemorial – still needing to evolve out of it.

It is entirely possible that Venus did once possess an agreeable climate that was subsequently destroyed by a runaway greenhouse phenomenon – and if this is the case, then something would have had to have caused it. There are some amazing revelations in ancient accounts and legends regarding the notion of life on other worlds. The story of Orpheus is one such account; Orpheus was the son of Apollo and the following fragment is attributed to him: "Those innumerable souls, they fall from planet to planet and in the abyss of space, lament the home they have forgotten." Does this not support the possibility that we have just reflected upon and suggest the migration of intelligent beings from elsewhere?

Heraclitus and all the disciples of Pythagoras considered each star as being the centre of a planetary system. This is an example of amazingly advanced thinking for 600 B.C. These ancient thinkers spoke of these things as though they were fact and not just theories generally circulating in their time. As I have speculated in other work, it is true that the ancient Greeks were receptive to wisdom and advanced thinking, and not fearful of it as some races were. The Greeks no doubt inherited this wisdom and knowledge from the ancient Egyptian accounts and historical records. So in turn we might then ask, who educated the ancient Egyptians? There is no escaping the indication that a vastly earlier epoch may have existed – an epoch where modern advances in technology, philosophical thinking and construction were prevalent. But today we live in times during which our academic establishments, and certainly most of our historians, are reticent to accept such an idea.

The ancient Greeks researched, preserved, logged, and expounded upon previous knowledge. They had a propensity for sound logic that is expressed in, for example, the statement: "To call the Earth the only inhabited world is as unwise as to assert that there is only one spike of grain growing in a cornfield." The Romans in their turn, who despite doing their share of conquering and pillaging, also had many enlightened ones who continued in the work of their highly respected Greek predecessors; instead of deriding their forebearer's enlightened remarks they often enlarged upon them.

For example, Lucretius (98–55 B.C.) stated that, "It is in the highest degree unlikely that this Earth is the only one to have been created." Lucretius was evidently not alone in his thinking at the time as seen in this statement by Cicero (106–43 B.C.): "The realm of Heaven is peopled by a host of other creatures." We have to ask whether the Greeks and Romans were as mystified as ourselves regarding UFO phenomena – it was likely that they were because they wrote about what they described as "flying shields". And as we know that the Romans worshipped many different gods that governed various aspects of their lives, perhaps we might consider the idea that their "gods" were actually extra-terrestrial beings.

Virgil recorded in his *Fourth Eclogue*, "Now a new race 'descends' from the celestial realm." It is difficult, when viewed through a modern-day lens, to see this in any other way than a cosmic visitation; so who were this "new race" of people? The Vedas of India were quite definite in their writings on life on other celestial bodies far from Earth. Perhaps their ultimate aim was to attempt to reach them in their flying palaces, or "Vimanas". Clearly, these writings of the Romans and the ancient Indians shows that they had similar wonderings to us about the possibility of extra-terrestrial life. But some of these ancient quotations appear to indicate they had already received proof of it, and that the concept of other worlds inhabited by intelligent beings was already accepted.

The ancient Chinese were also clearly interested in the possibility of extra-terrestrial life, but perhaps unlike in India, the Middle East, and so forth, had not yet witnessed clear proof in their skies. Nevertheless, the belief was well established as is evident in the statement of Teng Mu, a scholar of the Sung Dynasty, who summed up the view of ancient Chinese thinkers when he said: "How unreasonable it would be to suppose that besides the Earth that we know there would be no other Earths in existence."

The ancient Chinese were interested in most things celestial, such as the natural occurrences of eclipses and comets. Total eclipses are rare but there are many partial eclipses to observe. Comets on the other hand are more few and far between and the study of them requires a long history of observations. The ancient Chinese kept meticulous records of the comets – and their knowledge of them is not so far behind our own because comets still represent something of a mystery to us today. There is still no hard and fast explanation for them that is considered established fact. In old times, comets were feared by the more primitive races who considered them portents of doom. This is easy to imagine – especially if a comet was observable quite coincidentally during a major upheaval or dramatic event such as a flood, or earthquake, or even a meteor strike. Comets do not shoot by like meteors; it takes time from their first appearance orbiting the Sun to their gradual disappearance. Therefore, if a memorable or recordable event did occur when a comet became visible, then ancient people would naturally have feared its return.

The ancients were right to fear the comets as indeed we do, or should do, because we have seen the destruction that they can cause – for example in the case of the comet that exploded over the Tunguska region in Russia in 1908 and flattened trees over an area of more than 2,000 km^2. If this had occurred over a city it would have been even more destructive. But even more fatal would have been the one that was broken up by Jupiter's gravity and that subsequently made a series of impacts around its surface – any one of which would have been enough to put us back into the Stone Age if it had struck Earth. Fortunately for us, we have the edge over the ancients in that, if necessary, we possess the technology to go and land on comets (as we have done) at which point we could plant a nuclear device in order prevent its impact with Earth.

The tenuous composition of a comet makes them far less of a threat than a mountain-sized rock such as some of the asteroids that might come our way could be described as. Comets have been described as rather like a dirty snowball – they dissipate relatively quickly and shed their material in an explosive force that causes immense blast damage, but not a crater; although there is always the possibility that a violent meteor shower may follow on behind a comet causing further damage.

Scientists now think that in the distant past, in our own solar system, a violent catastrophe destroyed a planet and its atmosphere, and it is the remains of this planet that caused the creation of the asteroid belt. This would account for some of them, but there are masses of comets out beyond the planets, and collectively these are known as the Oort Cloud. To date, there is no sound explanation for the Oort Cloud's presence.

The creation of planets is just as mysterious as their destruction. The ancient book of the Guatemalan Maya, the *Popol Vuh*, contains the following quotation: "Like the mist, like a cloud and like a cloud of dust, was the creation." And this seems quite an accurate description of what science generally agrees may have prevailed before the so-called "Big Bang". We could also mention the ancient Chinese writings in the *Huai Nan Tzu* (120 B.C.) that stated that "Worlds are thus created by the crystallisation of pools of matter" – that is, by the gravitational condensing of a mass of cosmic dust, gradually contracting towards solidity.

Around the same time as the writing of the *Huai Nan Tzu*, Pythagoras referred to "the music of the spheres" – which wasn't a reference to music as we know it, but rather to sound given off in varying frequencies that the astronomers and astrophysicists of today (operating their radio telescopes) are well aware of. Pythagoras was perfectly correct when he stated that "everything that is and everything that is in space has a sound", that is, a frequency. People who control radio telescopes would certainly agree; it is a fact that the stars and the planets are all sources of radio emission in the form of sounds, and all have varying frequencies. This may also serve to indicate the problems facing the team of astronomers that make up the group known as the S.E.T.I Institute; with all the trillions of stars all emitting different frequencies, where do they start when searching for intelligent transmissions?

In the early stages of the S.E.T.I space programme, the pragmatic view was that since almost everything in the makeup of the universe was, and is, made up of the element hydrogen, then extra-terrestrials would most likely use this frequency for transmit on. And maybe these extra-terrestrials would also assume that any intelligently advanced recipients would also expect hydrogen to be utilised. So, the analysis of this frequency has begun but, as far as we know, we have not yet received any greetings from another world. I purposely state "as far as we know" simply because if the authorities are so reticent to release any positive information regarding UFO encounters, sightings and alleged abductions, then they may also be equally reticent to release any information regarding messages received. We are not in a position to know whether or not the members of the S.E.T.I have been warned that they must (in order to keep their jobs) first inform the secret agencies in the event of a signal rather than shout it out to the nation. The excuse, of course, is that announcing that extra-terrestrials have called out to Earth without first preparing the general public for such news, would cause too much of a panic or culture shock. It is true that a phase of social disorientation might occur, but that is another issue.

With regard to the subject of this chapter, that is space oddities, we must consider the strangest force that affects everything in the universe; even light itself, which is of course gravity. There is no question that gravity rules in the universe, but the most powerful of the four fundamental forces is that which is locked up in the atom. Electromagnetic force is much weaker but gravity is the weakest force of all – although this may be difficult to convince to a mountain climber who may have had a close call.

To insulate things (such as spacecraft or ourselves) against the force of gravity has been a constant dream of science and the subject of numerous science-fiction stories; and this may have actually been achieved by our mysterious visitors who allegedly periodically visit our air space. Intense research has been, and is, taking place into the quest to overcome it. When, or if, we do, the usual question will no doubt be asked as to why we did not realise it sooner; but as it stands (taking the pessimistic view), it may be a question of wonderment that will still be occupying our descendants for some time to come. However, if it is true, as many ufologists claim, that secretive research is being carried out on captured alien craft, then it is for certain that antigravity capability would be the first thing that investigators would wish to discover.

However, once again we find that our ancient predecessors were just as curious and interested in the subject as we are today. We have found that it is risky (through past experience) to say that certain things never happened or could never happen and write them off as sheer nonsense. A perfect example of this is the oft-quoted academic of a French establishment who stated with regard to meteorites "Don't be ridiculous there are no stones in the sky, so how can stones fall from the sky"? At the time, his logic may have seemed irrefutable with few disputing it, but already many centuries before him, people were quite aware of them. With regard to gravity, Françoise Lenormant wrote in his book *Chaldean Magic* (1877) that, "By means of sounds, the priests of ancient Babylon were able to raise into the air heavy stones that a thousand men could not have lifted." Could it be that this method was employed by the ancients to lift the massive foundation stones of the Temple of Baalbek whose massive dimensions defy all logic as to how they could possibly have been transported let alone been lifted into position?

It is a fact that gravity is a comparatively weak force, but we also know that it varies with mass. The dim companion of the bright star Sirius is composed of matter that is so dense that a mere cupful of it would weigh twelve tons – yet even this does not break any records. There are certain small stars in the constellation known as Cassiopeia where it is said that a cupful would weigh a staggering five million tons. These odd considerations are helping us on our way toward greater understanding of black holes. Matter cannot go on bearing such an intense force indefinitely and must eventually implode into itself. So where does this matter go?

Today, scientists believe that the occurrence of black holes can be explained, but so far, this is as far as we can go. We repeat, where does the matter go? Some things may never have an explanation – like, for example, trying to imagine eternity. Or, trying to imagine an object travelling on forever never stopping if nothing exists in the realm it travels in to stop it, for example outside the farthest boundaries of the universe, if indeed the universe has a boundary. The concept of the "big bang" after all implies that at least at one time, such an empty space existed.

If we return to the lesser mystery of the temple of Baalbek, the base stones are so massive that it seems extremely unlikely that slave labour was used to accomplish the task; especially considering the fact that no massive cranes were available. Naturally, the same set of questions arises when the building of other huge, gravity-defying edifices is considered, such as the Egyptian pyramids or even Stonehenge. Of course, in the case of the pyramids, it is not so much the size of the stones, although formidable, but rather the rate at which they would have to have been put in place in order to make sense of the short period of time that it apparently took to build them. And again, mythology steps in by offering strange and hard-to-accept theories bordering more on magic than logic; tables from Arabic sources state that the stones for the pyramids were wrapped in papyrus, and then struck with a rod by a priest which made them weightless and easy to manoeuvre into place.

It is said that the drawings, paintings and depictions carved into stone of masses of labourers hauling the stones up ramps, and females pouring jugs of water before the stones to make the sand beneath ready, were all produced long after the edifices where erected and were simply based on the logical conclusions of the artists as to how it was accomplished. If we refer again to the rate at which the stones would have had to have been put into place, it was ascertained that each stone would have to have been manoeuvred into position every two and a half minutes to accomplish the building of a pyramid such as the Khufu Pyramid in the time allotted.

In *Pillars of Fire*, I wrote about the destruction of the walls of Jericho – a feat that was accomplished by the use of concentrated sound during the adventures of the Hebrews. It seems in this case, that it was the frequency of the sound waves that provided the force.

Oddities of Earth and Cosmos

To return to the lifting of great stones the Copts who are alleged to be the direct descendants of the ancient Egyptians, state in their writings that the blocks were raised by the sound of "chanting". Or perhaps this would be better translated as "shouting" because mass "shouting", along with the sound of "trumpets", was necessary to destroy the walls of Jericho (as previously mentioned). So sound is definitely a force, but as far as modern science is aware, there has not yet been a discovery about sound having the ability to achieve weightlessness.

Lucien, of the second century A.D., spoke of antigravity as a reality during his time. He testified on an event that he himself witnessed in a temple in Hierapolis, Syria; he stated that he witnessed the god Apollo "leave the floor where the priests stood and was born aloft." When the Queen of Sheba travelled to Israel to see for herself the wonders of Solomon and the alleged achievements that had reached her ears, her wishes were duly granted as she personally witnessed Solomon rise up from the ground in full view of her. At this, all her scepticism and suspicions of exaggeration disappeared, so much so that, as the Bible puts it in the Book of Kings, "There was no more spirit in her."

There are many periods listed in ancient history where references are made to antigravity and to the ancient adepts who had scientific knowledge and skill not yet without our own, modern, grasp. The twelfth century historian and cleric, Geoffrey of Monmouth, mentions in his work *De gestis Britonum* (or, *Historia regum Britanniae*), the legend of how the huge rocks precisely shaped and fitted in the structure of the ever-mysterious Stonehenge came about. His story relates that the legendary figure, Uther Pendragon, who was the father of the equally well-known King Arthur, led a substantial force of 15,000 Britons, who at the time occupied the area where the stones for the great historical assembly were to be placed. Soon after they had secured the land, they then set themselves the task of removing the boulders. At this point in the story, we must assume that the learned monk meant the quarrying site because he used the term "removed" rather than "assembled".

It is established that many of the stones were cut from the area of the Preseli Mountains in Wales – a location that is quite a distance from Stonehenge and only the first problem to that needs to be overcome. To continue the account related by the good Monk – the large amount of people who were involved in the building project were manfully struggling to transport the stones by using all the conventional methods available at the time such as hawsers, ropes, scaling ladders and so-forth, but the army of men "could never a whit the forwarder, take the stones."

And then, quite unexpectedly, on hearing hilarious amusement, they all stood still wondering who might be mocking them and their efforts – and they soon realised that the laughter was emanating from Merlin the Wizard who then came forward indicating to the men that they should stand aside. A monk named Gerald described it as follows: "He [Merlin] began to put together his own "engines" with which he lifted and put the stones down [into place?] so lightly, that none of the people could believe it possible." Whatever these special "engines" were, remains open to our conjecture, but we may assume, that Merlin was well acquainted with all forms of leverage and mathematical engineering processes that we may well recognise today, but that would have astounded the ancients.

Nevertheless, whatever the method was, Gerald made it clear that "by means of these special engines the stones were transported and set up at Stonehenge" and he continued, noting that this fact "proved yet once again how skills surpasseth strength." King Arthur's army of men who made up the workforce must have been responsible for some of the ancient stone masonry and the clever forming of the notches and protruding pillars that enabled the stones to be so precisely fitted together – even more precise considering that the whole edifice is built on sloping ground.

However, we are not to know where Merlin's assistance stopped or how much more he was responsible for. It is a pity that Geoffrey of Monmouth did not further enlighten us as to what the "special engines" were. Perhaps a similar process was used in ancient Egypt at the time of the pyramid builders. We have mentioned that just as we begin to wonder if the builders of ancient times did possess methods bordering on magic to overcome gravity, we encounter some stories that make us consider whether they are just elaborate stories and nothing else.

The following story about the possibility of overcoming gravity is a good example of this and can only be described as rather silly. According to the biography of Liu An contained in the *Shen Hsien Chuan* (fourth century A.D.) he could swallow a certain liquid concoction, or 'elixir', that would make him become suddenly airborne. This in itself is hard to believe, but it gets stranger. After leaving the container standing in the courtyard, it was knocked over and consumed by a dog and various chickens – and so soon, they too "sailed up to heaven" and the sound of dogs barking and cocks crowing could be heard in the sky. Perhaps this is just an example of ancient Chinese humour.

Nevertheless, there are some more seriously related accounts that perhaps some of us might take as read when, for example, the Bible tells us in the Book of Kings that Solomon during his relationship with the Queen of Sheba "rose up in full glory." It is tempting to wonder whether there could be a scientific explanation for such accounts of people shielding themselves from the force of gravity. But the easier response would be to simply write them off as nonsense. Yet let us not forget the most famous of all examples which is of course when Jesus himself was said to rise up into the clouds in full view of the disciples.

We can also quote further examples such as the story of Simon the Magus ("magus" meaning magician) who was said to have been a first century Gnostic philosopher. It was written, that this "magi" was able to perform miracles by means of his magical science; one of the stories relating to him was that when he was addressing a crowd of thousands in Rome, "the spirits of the air" helped him to rise high above the ground. He was also said to be able to raise heavy statues by causing them to lose their weight and glide through the air.

The Catholic Church seem to need no convincing of the reality of levitational feats by humans, as it lists some two hundred saints who were alleged to have overcome the force of gravity. Emanating from religious sources no doubt gives these stories some credence, but two hundred of them is quite a substantial number. One of these accounts tells the story of St Benedict (A.D. 448–548) who, when in Mount Casino in Italy, was said to have lifted a large stone that was extremely heavy in weight by neutralising gravity. When the stonemasons of the day could not lift it, St Benedict made the sign of the cross on the rock and lo, while the group of stonemasons looked on in amazement, he raised it aloft without any effort or help and put it neatly into place. Just as I wrote about in *Strange Realities* we could not expect everyone to believe amazing accounts such as these – or as related, for example, to the activities of the Count de St Germain. If all the extraordinary events and achievements had not been witnessed by people who supported their authenticity by writing about them in their memoirs, it would be easy to dismiss them all as fiction.

The following example is a case in point, it has to be said that it does stretch our credibility somewhat, but the important point is, that it was witnessed by the Spanish Ambassador to the Papal Court. It appears that in 1645, ten men were struggling to lift a large eleven-metre cross into a pre-prepared position, and they weren't having much success. At this point, St Joseph of Cupertino (1605–1665) moved sixty metres through the air, picked up the cross with ease, and installed it into its position. And this occurred, as mentioned, in the presence of the Spanish Ambassador, his wife, and indeed the whole congregation of the church who were said to be completely spellbound and astonished by the event.

During the time of the British Raj in India, many servicemen, officers and soldiers witnessed many astounding events carried out by Fakirs and Indian magicians – with feats of antigravity being among them. But even though these servicemen were astounded, they put them down to tricks and illusions. If they had instead professed their belief in them, they would probably have found themselves on the next troop ship home being diagnosed with heat stroke and relieved of their post.

Some ancient accounts regarding the concept of gravity and its effects, take the view that the force is variable and not a constant. In a sense, this is true because it is said that at various points around the globe – from the arctic, to the equator, and onward to the Antarctic – the gravitational attraction does indeed vary.

The amazing cases that we have relayed so far are mostly events that took place in classical or medieval times. However, there are more cases of these feats that occurred in fairly recent times and that were verified by actual known individuals from history. For example, in a letter dated 14 July 1871, Lord Lindsay, recounted a strange experience that he had witnessed in the residence of a certain Daniel Dunglas Home. In his letter he wrote:

"I was sitting with Mr Home and also in the company of Lord Adare and a further guest who was a cousin of his. During the sitting, Mr Home went into a trance and in that state was carried out of the window, we all witnessed Mr Home floating in the air outside our window. This occurred at 5 Buckingham Gate. He remained quite motionless in that position for a few minutes and then began to glide back into the room, feet foremost, he then sat down".

The window that Lord Lindsay was referring to, was seventy feet above the ground. One would have thought that if D. D. Home was prone to such spontaneous (one might call) "attacks", then he surely would have been a lot safer if he had sought an apartment on a much lower level, or at the very least have kept his window closed. If it was a form of somnambulism or sleep walking (or in his case sleep "gliding"), it would have been even more dangerous since if he had been woken by someone shouting in sudden alarm, it may have woken him out of his semiconscious condition and caused him to fall.

Despite the potential danger, this was not the only occasion when D. D. Home was witnessed to levitate by historically known figures. Further confirmation of D. D. Home's strange ability was confirmed by the noted British Physicist Sir William Crookes who played a part in the eventual development of television with the invention of his "Crookes Tube". Sir William watched a performance of D. D. Home levitating on three separate occasions – a fact that he confirmed in 1874.

Astronauts most certainly know what it feels like to experience a lack of gravity. Early on in their education programme they undergo a training whereby their aircraft climbs, and then descends after performing an ark-like manoeuvre. At this point, the crew are briefly lifted off the cabin floor because another force has taken over, equal to gravity, and caused them to defy it. However, the aircraft and its occupants are still falling toward Earth and are therefore, even in these moments, still subject to gravity although it doesn't feel like it.

A skydiver in freefall feels weightless and can freely gyrate about, just as the astronauts in earthly orbit, but the skydiver and the astronauts are still falling. The astronauts are falling around the Earth, but do not make contact with it because their orbital speed prevents it – this occurs at around 17,000 mph. We could put this in easy terms by imagining a person throwing a cricket ball – it will travel forward at first and then sink to Earth, but if a taller, stronger person threw the ball then it would travel a bit further, and then if some superhuman came along and was able to throw the ball hard enough, it would not sink to the ground at all but simply keep spinning, or orbiting, around the Earth until its velocity was impaired. So in the case of our astronauts, a simple forward thrust or "burn" is needed.

When astronauts increase their velocity to 25,000 mph, they totally overcome the Earth's gravity and move toward the Moon until another gravitational force, i.e. that of the Moon, begins the process all over again. However, levitation is a different phenomenon and defies all known laws of physics. The Bible tells us that Jesus did this. Who are we to compare ourselves to him, yet Jesus did say "What I can do, ye also can do." Does this mean that superhuman powers are latent in all of us?

Evidently only a tiny number of extraordinary figures such as King Solomon and others achieve this kind of feat in reality.

One of the most profound mysteries of our time is to imagine that a culturally-advanced human race lived on Earth millions of years ago. But this is exactly what the ooparts (such as various machine-cut metal objects) that have been found in lumps of rock and coal, imply. We can offer a rather weak number of possible explanations but none ever comes close to a valid explanation. So-called "New Age" thinking would suggest that an extra-terrestrial race lived on Earth millions of years ago. But this explanation would be as unpalatable to science as the alternative – and so it seems that the scientific community have preferred to take the easier path and just pretend that these objects don't exist, or ignore them. The very best-case scenario in these cases is that scientists defer the responsibility of proposing theories and just say that a logical explanation will emerge at some point in the future. To bravely confront these mysteries, and to consider the possible implications of the fact that such an ancient technology may once have existed, would be to open a Pandora's Box – it would require the complete rethink of all we thought we knew about the history of humanity. But, it might also help to explain many other seemingly unexplainable phenomena and it would no doubt contribute to our own further technological advancement.

To ignore all the detailed writings and descriptions of flying craft delivering fearsome nuclear bombs on the heads of their enemies in the Hindu epics of 4,500 years ago, would have been an awful waste of expensive ancient parchment if it had only been for entertainment. In any case, these were cultures that passed their stories on mainly through oral, rather than literary, culture – so why write these things down in the first place? These epics were clearly written to preserve ancient Hindu history and to recount the technological capabilities of their time. And, the technology itself would have indicated a long history because it would have taken a vast amount of time to develop – just as the technology that, for example, was used to build the ancient megaliths would surely also have taken a long time to develop.

If we imagine that technology existed millions of years ago, it alters everything. If it is true, then what else was happening in our solar system at the time? In our own day, mobile exploration units are trundling about on Mars and they have often encountered objects and edifices that have aroused enormous speculation; for example, possibly constructed pyramids, an alleged "face" that is said to be bilaterally symmetrical, and even a suggested sphinx. Some of these discoveries (but some would suggest not all) have been revealed to us; one of the most important discoveries is that water once flowed in copious amounts on Mars and that it could have been a world that was once occupied by living entities (perhaps even by ourselves).

And what of its neighbours? Perhaps Mars once had a planetary neighbour that existed where the asteroid belt now exists? Planets do not explode – so an outstanding occurrence may have brought about a possible accidental destruction of the world, a war in Heaven if you like. Imagine an atomic missile plunging down through the magma of a volcano (in Hawaii for example) and then exploding violently underneath the crust. An event of this kind may cause a planetary disturbance such as a violent wobble, causing other forces to come into play and wrench the planet apart. Could this have happened with regard to the hypothetical planetary neighbour of Mars?

The massive planetary gravitational forces of Jupiter may have reduced the outer layers to orbiting rubble and shifted the planetary core onto a new track. In *When the Moon Came*, I speculated on the possibility that all the forces of celestial mechanics, coincidentally favourable for a capture and also solving other currently unexplained mysteries, assured our Moon's arrival in Earth space.

Of course, if this catastrophic event did occur, Mars may also have suffered greatly; perhaps by having had its atmosphere torn away by all the horrifying events that brought about the conditions we see on Mars today. The rises and falls in the advancement of human technology that we know took place in the past, may go back far into distant epochs. And perhaps some of the artifacts embedded in stone and coal were a result of construction and manufacturing that was occurring then. So, an acceptance of this distant technology having once existed would explain, as we have depicted, many other things that have puzzled us in the past and still continue to do.

Examples of these mysteries include: human boot prints (with a ribbed sole and stitching) in ancient sandstone; human footprints being found together with dinosaur tracks as if the human was stalking the creature; a fossilised human brain from the carboniferous period found in Russia; an ornamental silver bell-shaped vase found under fifteen foot of solid rock during blasting operations; a gold chain found in a coal system. The list is lengthy, but who is brave enough to broach these things and turn history on its head? Do we just walk away from it all? Perhaps the scientific community have decided that they'll have more restful nights by taking this option.

CHAPTER VIII

ANCIENT SCIENCE

We cannot help going back to the thought of Solomon; as we have indicated, almost everything of a scientific or technical nature that we take for granted today seems to have its ancient counterpart: automatons, computers, dispensing machines, calculators, and robots who could serve food and communicate verbally.

It is written, that the Egyptian temples had slot machines that dispensed "holy water". They were so cleverly and precisely designed that whatever coin was put into the machine, the amount of water dispensed was proportional to the weight of the coin inserted into the slot. The ancient Greeks, in their legends, had a plethora of "gods". Among them, was one important god who was known as the "blacksmith of Olympus" who was constantly hammering away busily producing his marvels of mechanical construction in his mountain smithy. Some of the Greek legends and indeed the legends of many countries, when we encounter them in written form, may seem unacceptable – but while the core of the stories may have sound authenticity, we cannot ignore the possibility that they were embellished before being written into legend.

With regard to the Olympian blacksmith, it is written that he made two golden statues in the form of young women – and these statues were apparently frequently by his side, assisting and serving him. And outside of Greece, many stories of automatons and robotic entities also show up in the historical legends of other countries.

The indications seem to imply that what we would call "cybernetics" seem also to have been a product of ancient science. As I mentioned in *Pillars of Fire*, these stories also appear in Chinese and Tibetan legend. The Greek philosopher, Apollonius of Tyana, was said to have seen automata and dumb waiters when he travelled in Tibet. And in ancient Chinese legend, the story of the Emperor Ta-Chouan is a good example of the kind of automaton that crops up. It is a fascinating story, although the fact that it goes into more detail than most other stories of a similar nature makes one wonder how much it might have been embellished. It features the construction of a mechanical man. This automaton was said to have appeared so attractive to the wife of the emperor, that she became infatuated with it and paid more attention to it than to the emperor himself. It must have been extremely life-like because as a result of the attention that his wife paid it, the emperor became extremely jealous, and in a fit of rage he ordered it to be smashed to pieces.

China was certainly at the forefront with regard to ancient inventions. It invented the first calculator in the form of the abacus which was in common use 2,600 years ago, and is still used today. When using this device, calculations could be made so fast that Chinese bookkeepers used it widely. Yet, for all this, the most efficient of computers, both ancient and modern, is the human brain. Constructed devices of today are capable of wondrous things and capability, but they were all envisaged, pre-planned and designed by humans.

In 1959, when he lived in Sydney, Australia, the author Andrew Tomas witnessed an interesting event. At that time, the Indian Shakuntala Devi was known as the "human calculator", and she was asked by the University (she was likely studying there) to face "UTECOM" – Australia's greatest electronic brain. In 1959, before modern miniaturisation, UTECOM was a huge machine, but today it would probably fit into one's pocket.

Devi was asked to find the cube root of 697, 628, 098, and 909. In around six or seven seconds, Devi gave the answer, 8869. The electronic machine produced a slightly different result. However, Miss Devi confidently answered, "I'll bet my answer is correct" and, sure enough, after a series of cross checks and further computations with the machine, the new result showed that she was indeed correct.

The complexities and immense capabilities of the human brain are known about but largely taken for granted – and so they deserve deeper analytical thought. Where did these qualities come from? Why are humans the only beings in this earth with these gifts? To develop naturally, they would surely have required an enormous amount of time that anthropology does not allow for. If anthropology allowed for enough time for this kind of development to occur, then their standard chronology of human development would be thrown into disarray. Either way, whether they occurred naturally or were bestowed upon humans, we might wonder why these amazing abilities are not distributed more widely among humankind. If certain humans are capable of them, then theoretically, all humans should be equipped with them. So why do these capabilities only show up in a few individuals, and moreover, only show up in a set of humans who are so widely spread across the entire world? And, if anthropology is aware of the super-human feats, why does it cling to the outmoded assumption that such qualities are inherited from other primates as Charles Darwin proposed. And if Darwin's theory has not yet been proven, then why continue wasting so much time trying to make the apes "fit the bill" so to speak, by practically forcing human traits upon them in training programmes? This is a question for a different book, but there is no doubt that the ancients used their brains to their fullest capacity.

But to return to the subject of engineering and mechanical capabilities, the engineers of Alexandria in Egypt constructed over a hundred difference types of automatons, or mechanical men, over 2,000 years ago. In Greek legend, Daedalus was a prolific inventor. Unfortunately however, his son, Icarus, came to a sticky end when he flew "too close to the sun," as the legend goes, when piloting one of his father's inventions. Daedalus is also said to have constructed human-like figures that were able to move on their own accord. Plato, who we must assume would not be easily fooled, referred to these robots of Daedalus', stating that they were so active that they had to be quickly restrained in order to prevent them from running away. We can only wonder at the mechanical arrangement that would have been capable of such a thing.

Ancient Science

Perhaps this would be a good point to mention the ancient artifact, about the size of a portable typewriter, that was found by sponge divers off the coast of the Greek island of Antikythera. When it was x-rayed, it was found to contain an intricate mechanism consisting of cogs and wheels of intermeshing gears that must have required the skills of a watchmaker to construct. It could have been utilized in many ways – for example for navigation, star fixing, the prediction of astronomical events, and it was perhaps capable of all the tasks that a marine chronometer fulfils. It is thought that it may have been constructed in the second century B.C. and so the intricacy of its design and manufacture are astounding. It is no wonder that one scientist, when commenting upon the discovery, said that it "was like finding a jet plane in Tutankhamen's tomb."

Ancient Egyptian texts tell us that in the Temple of Thebes, Egypt, there were figures of gods that could speak and make various gestures. These must have been robot-like constructions that could move independently, because it was assumed by some witnesses, who could not imagine any other way it was possible, that the priests must be hiding inside and manipulating them. Evidently, the ancients that witnessed these "gods" were quite wise and not easily taken in.

This was quite unlike the situation during the time of the British Raj when the British tried to convince the Indian natives that the magic performed by their fakirs and travelling magicians was all just trickery. The British warned the natives that they shouldn't be taken in, or exploited, by these tricksters for the few rupees they had. Yet, for all the British confidence at that time, can anyone even today explain how anyone can perform the Indian rope trick out in the open air? The British confidence was likely only skin deep in the face of such mysteries.

And the same might be said of the conquistadores who first encountered the Incan natives in South America. Although the story goes that the Incans were impressed by the invading Spanish, the Spanish must have been puzzled by many of the things that they witnessed. They must have wondered why there were proficiently-constructed roads all the way down along the South American west coast from Ecuador to Chile when the Incans didn't utilise wheeled vehicles. And they also must have been awestruck by the effigies that were said to mimic human-like activity. According to Garcilaso de la Vega, a Spanish soldier and poet (1501–1537), the Incas where in possession of a statue that spoke and gave answers to questions. It was situated in the Valley of Rimas.

And this story of talking effigies echoes those that come out of Greece and Egypt – for example the oracle of Amun-Ra in ancient Egypt. The Amun-Ra was an automaton as well as an oracle and built in the likeness of a god. This effigy could not only walk, talk and move its head, it also accepted scrolls with questions on them to which it gave intelligent answers. Even the worldly-wise Alexander the Great encountered it. It is said that the figure came forward to meet him. It is not recorded whether Alexander put in a question on a scroll or asked a question directly, but apparent the effigy spoke to him and said, "I give to thee all of the nations under they feet."

The makers of the film "Jason of the Argonauts" was inspired by these ancient Greek and Egyptian tales. It was the great special-effects wizard, Ray Harryhausen, who was employed to make the film come to life. He created the figure of the god "Talos" for the film. This figure knelt on one knee on a plinth with a sword in his hand, and in its efforts to slay the Argonauts, it came off its plinth and tried to destroy them. We have heard of the expression "Achilles heel", but in this case of the film, it was Talos who had his heel as his weak point. When Jason unscrewed a stopper in Talos' heel, all Talos' "ichor" (or blood of the gods) flows out and Talos falls to the floor and shatters into pieces. This is similar to the original Greek myth – the difference being that the weak point was Talos' ankle in the original, and the slayer of Talos was Medea rather than Jason. The Greeks were great tellers of tales, so we have to attempt to separate what might be real events from fiction. But it is curious to note that the Greeks were not the only ancient civilisation to envision robots and include them in their myths. In the original Greek myth, the job of Talos was to protect the island of Crete from intruders. The message of the film echoed that of the Greek myth – that however powerful you are, you will always have a weak point, or Achilles' heel, or Talos' heel, that can be exploited. According to Greek legend, Talos was the last of a race of bronze humans. This idea is strangely futuristic as it has actually been envisaged that at some point, perhaps not too far ahead, virtually indestructible robotic "terminators", mostly of metal, may exist to do the "dirty work" of the infantry.

The idea of robotic beings can also be found in ciphered medieval books on magic that were preserved for centuries. The monk Gerbert D'Aurillac (c.920-1003) who became a professor at the University of Rheims, and later Pope Sylvester II, was reported to have possessed a bronze automaton that also answered questions. It was constructed by Gerbert and answered "yes" or "no" on important matters concerning politics and religion. It is likely that Gerbert knew only too well what each answer would be because the answers would have been pre-programmed by him. This story makes one wonder what kind of other stranger and mysterious stories are contained in the vast collection of works in the Vatican library. Will the library ever reveal its secrets? Will scholars ever be allowed access to it all? Surely there would be little point in preserving such a vast collection of books if we can never learn anything from them?

The Bishop of Regensburg, Albertus Magnus (1206–1280), was a very learned man. He wrote extensively on such subjects such as chemistry, medicine, mathematics and even astronomy, and it seems that he too was interested in robotics and constructing automatons. According to his biographer, it took Magnus over twenty years to build his famous android. The biographer goes on to say that his automaton was composed of metals and unknown substances. The robot was, like for example the ancient Tibetan automaton, also said to have been capable of walking, talking and carrying out a range of domestic chores.

The famous philosopher and theologian, Thomas Aquinas, was a disciple of Albertus Magnus; the two lived together and the automaton looked after them both. As Thomas worked at home during the day, it seems that it was he who had to bear the brunt of the robot's strange and constant chatter. The story goes that the constantly chatting robot eventually drove Thomas Aquinas to distraction, and so one day he lost his cool, took a hammer to it, and smashed it up. All this sounds very much like the aforementioned story of the Chinese Emperor, his wife, and his robot, and like the Chinese Emperor, Aquinas ended up despising the machine. It is likely however, that after taking twenty years to build it, Albertus Magnus was not too happy with the incident.

It would be easy to dismiss all of this as entertaining legend, but Thomas Aquinas was a well-renowned scholar. He was learned enough to, for example, recognise that the Milky Way was a dense conglomeration of distant stars (although in truth we now know that in actuality, the stars are quite spread out in a circular fashion and it is simply due to the fact that we view the galaxy edge-on that that the stars appear densely placed). The writings do not tell us how Albertus Magnus reacted to Thomas Aquinas after he destroyed his robot, but in any event, both of these men were made saints by the Catholic church – so Aquinas' violent outburst did not jeopardise his ecclesiastical promotion (and proves that even saints lose their temper!).

With regard to all these amazing mechanical inventions, we have two options to consider – they are either myth or truth. It is easy to accept them as myth because myths exist everywhere, but to accept them as truth we have to have proof and this we do not have; so myth prevails until archaeologists unearth evidence. What we can be sure of is that the concept of mechanics and the manipulation of metal parts obviously existed at least in the minds of the ancients. And, the skill to produce functioning cogs, gears and metal parts is clearly evident in the Antikythera object (mentioned earlier) that was found by the sponge divers off the coast of the Greek island.

Ancient Science

Another pertinent factor to consider, is that many of those who claimed to have witnessed these wondrous objects were not obscure individuals but were often well-known figures in history with their reputations to lose. For example, I mentioned Apollonius of Tyana (who was covered in my book *Pillars of fire*) who when in Tibet encountered robotic entities that moved of their own accord and were "subservient to the beck of the gods." These robots even delivered food to the table of the seated guests.

Conventional thinking has it that the seismograph was invented in 1703. Yet, the ancient Chinese had inventions and artifacts that fulfilled the same function. So, while Jean de Hautefeuille did build a seismograph in 1703, it is thought that the Chinese scholar Chang Heng (or, Zhang Heng, A.D. 78–139) also constructed a seismograph a millennium-and-a-half beforehand. Heng's seismograph was built in the form of a vase and decorated with dragon's heads holding balls in their mouths. Below each dragon's head was a porcelain frog. When an active earth tremor occurred (even if it was too far away for anyone to detect), the ball would fall from one of the dragons' heads and into the mouth of the corresponding frog. This would tell the seismographer the magnitude of the quake and the geographical direction in which it was taking place.

With such intricate ancient ingenuity – for example, the cleverly-constructed artifacts and the accuracy of astronomical calculations – we mention again how strange it is that all this knowledge was lost in the medieval period, only to resurface with the discoveries of Newton and others like him, during the Renaissance. We have provided many examples of this throughout this book. Although highly intelligent individuals obviously existed during this time, there was also some quite astounding ignorance. For example, about 150 years ago, at the World Expo in Paris in 1878, members of the French Academy of Sciences accused Thomas Edison of being a charlatan and a trickster when he demonstrated his phonograph (an early form of gramophone that could record and play back sound) to them. It seems incredible that alleged men of science could react in such a way without even examining, enquiring and seeing for themselves how it was constructed and how it operated (as would be the scientific approach). And this occurred in fairly modern times! It makes one wonder what kind of examining body would deem such men as qualified to become members of the Academy of Sciences.

And it seems likely that this type of technology was also in existence in ancient times; with all the automatons that have been mentioned in various texts, and their ability to talk articulately, the ancients must have had knowledge of how to manipulate sound. Perhaps all the ancient drawings and diagrams that contained all the necessary instructions needed to construct these figures, were purposely destroyed (like so many other similar records) through the ignorance of perpetrators who existed at the same time as those with knowledge.

In classical times, Roman historians in Egypt described the "singing statue of Memnon," which was erected in 1350 B.C. Musical sounds were said to emanate from the statue, which faced east. Apparently, it was in the mornings, when the rays of the rising sun fell on the head of the effigy, that it "sang". So in this example, we see a connection between the light and the heat of the Sun, with the production of sound. Perhaps built into the device, there was a delicate mechanism that only had to expand slightly to "set the wheels in motion" so to speak. But what of the recording device that must have been built into it? A singing voice must have been imprinted onto a kind of recording tape. So however this machine worked, it would have been extremely advanced technology for its time, and as usual in these cases, we are always left guessing.

But as well as being a sensitive piece of equipment, the recording device that existed inside the statue, must also have been quite robust because it was still in existence when the Roman Emperor, Hadrian, had the chance to listen to it one morning in A.D. 130. If this is true, it certainly indicates an extremely long existence for such a device that would have been subject to all the usual forces of deterioration, corrosion, and the condensation effects between night and morning in a hot climate etc. Is it possible that the inner workings of such a device could last for as long as nearly 1500 years? In fact, it would have been even longer because it was stated that the Roman Emperor Septimus Severus also heard the statue chant at dawn a few decades later at some point between A.D. 193 and A.D. 211. Such a long life seems more like the lifespan of a myth or legend than a physical device.

Emperor Hadrian actually encountered this effigy and its performance on three different occasions. Evidently, the stone that formed the exterior of the statue was more prone to corrosion than the inner workings that made the sound because apparently, whenever the statue's cracks and chips were repaired, the sounds that usually emanated from it stopped. Perhaps rough handling by the workmen, who surely should not have been trusted with protecting the delicate inner workings, was the cause. The statue of Memnon is still visited by tourists today although the sound can no longer be heard.

We return briefly to the Greek philosopher, Apollonius of Tyana, and his experiences in Tibet. In order to highlight the fact that there were a whole range of robots or mechanical devices in use in that region, we would mention that the units that served Apollonius and his group at their table were quite different from the human-like robots mentioned. It is written that these particular devices were tripod automatons. We can imagine their movements – they would have been at less risk of falling over than a bipedal construction (such as a human). And what additional devices might they have had (as we have arms and hands) in order to bring food, drink, utensils and so forth to the table?

Interestingly, this story of three-legged units is echoed in a fairly recent incident involving a "close encounter of the third kind." An apparently hitherto "down to earth" and no-nonsense man was employed as a forestry worker in Scotland. One day, as he was going about his business in the forest, he came to an abrupt halt when entering a clearing; there before him, was a semi-visible or opaque UFO, just above the ground and from it, there emerged a group of tripod or three legged units that began moving toward him. At that point, either the shock of it, or some device that the units employed, caused the woodsman to lose consciousness. It was probably the latter because being the type of character that he was, he would probably have raised his axe, gun, or whatever he had in order to defend himself. To verify the above encounter by the woodsman, investigators examined tears in his clothing and indentations in the ground of the clearing. The tears indicated that he had been lifted up by his trousers, and the indentations in the clearing indicated that something heavy had recently been there. Yet, there were no track marks leading to or from the area, and no heavy machinery matching the marks could be found anywhere nearby. But the point is, how do we explain the similarity of the robotic units witnessed in Scotland, to those that were "at the beck of the gods" some 2,000 years previously? It is interesting to note that Apollonius lived around the same as Jesus – and so some have suggested that they were one and the same person.

As I have mentioned in other work, there are some things that we just cannot improve upon, the wheel for example; it is always going to be round. When we think about Merlin and his "engines" in the building of Stonehenge, we might imagine that the devices he utilised so efficiently may have been instantly recognisable in regard to their purpose by our engineers of today – but seen as magic by the ancient labourers. Engineering always goes hand-in-hand with mathematics, which no doubt, Merlin had a good grasp of.

We still look upon Stonehenge with a certain amount of awe, as perhaps we should when we consider other factors besides looking at it as a group of stones arranged in an interesting manner. Obviously, quite modern types of sophisticated tools, instruments and measuring devices must have been used. They must have used devices such as a plum bob to establish verticality, as well as effective and functional surveying techniques and instruments. They must also have used sketches, drawings or some version of blueprint in order to plan the design and construction.

Ancient Science

This observatory of sorts was not erected on ideal level ground, and this factor would have made the construction more complicated than it otherwise would have been. The tilted ground would have had to have been accounted for when determining verticality and the correct levelling of the stones. They achieved this with an astonishing degree of accuracy; the top stones were, and still are, perfectly horizontal. And even more importantly, and the most difficult thing to achieve, was digging holes of the correct depth for the stones to be lowered into. The builders would even have had to allow for the compressive weight of the stones that when lowered into the holes would have compressed the soil down a further few inches. So if this had not been allowed-for, then the horizontal alignment of the capping stones would have been faulty.

As previously noted, modern-day builders and engineers would not be able to accomplish such a task without spirit levels, yard tape, plumb line and so forth. These ancient people must have had their equivalents. Did they have glass in ancient Britain? Simply filling a square bottle with water (but not absolutely full) would produce a crude version of a spirit level with a bubble of air in it. But they must have been in possession of more sophisticated equipment. They would have needed elevation sites and surveyor's equipment and accurate information with regard to the contours of the land, and the soil type and composition. Also, the stones differed in size and function, so the varying weight of the stones would also have had to have been calculated. Moreover, holes had to be bored into the vertical stones at a precise depth and width in order to exactly fit the protrusion of the proportionally-shaped top stones – to form type-of mortice-and-tenon joints in the stone. This would be a significantly difficult task to accomplish today, so the fact that it was accomplished so long ago is mind-boggling. And this was all accomplished by people who the Romans allowed the rest of the world to believe were "civilised" by them.

The building of Stonehenge occurred around 5000 years ago during times that we call "ancient". But it is all relative; how "ancient" is ancient?! The biblical King Solomon made it clear that most things had been achieved before his time; and he was not alone in assuming that the antediluvians had great knowledge. The Babylonian King Assurbanipal (remembered as being the last king of Assyria) was on record as saying that he loved to read of the accomplishments of his ancient predecessors that lived before the flood. King Assurbanipal was known to be a great supporter of culture, and he collected many old texts (texts that were ancient even at that time) in the form of clay and stone tablets – on topics such as astronomy, divination, medicine, as well as literary epics and myths – and it was he who was credited with the founding of the great library of Nineveh.

The antediluvians, whose births and deaths are all precisely recorded in the Old Testament, give us the ability to calculate that Adam, the so-called "first man", only appeared around 5,000 B.C. The fact that Adam was the "first man" is of course nonsense; mankind is far older than that. Adam should perhaps more accurately be described as the second man – a further creation by a god or gods who were perhaps not of this Earth. So before Adam, there existed a race of real antediluvians who have been named as the "pre-Adamites". The origins of humanity are complicated by the "second man" Adam, and we are still striving today to recover all the knowledge that went before him.

The discovery of the Ebla tablets in Syria in the ancient city of Tell Mardikh in 1975 caused a great wave of excitement at the time. They were found to have been inscribed in 2300 B.C. and were written in one of the Canaanite languages. These tablets numbered around 15,000 pieces and revealed a wealth of information that included political treaties, records, laws, religious texts and much more historical information – all of which tells us just how advanced and organised these ancient peoples really were.

However, as we know, even in the days of King Assurbanipal (around the seventh century B.C.), the knowledge of metals, alloying, the fabrication of metals, as well the knowledge of advanced mathematics, was known and would have been necessary to create machinery. And as we have seen, they most certainly did produce machinery – the evidence of which can be seen all over the world.

A prime example is the amazingly accurately cut and incised stonework evident in the ruins of Pumapunku in Bolivia. Experts have declared that the smoothly-incised angles, grooves, and circles could only have been achieved by a type of laser-cutting technology. At the site, the giant pieces of cut stone are scattered about here and there like a kind of mammoth jigsaw puzzle. They look as if they were either purposely pulled down, or wrecked by an earthquake. The latter possibility seems unlikely because so many other cleverly-constructed edifices at Pumapunku (that were evidently built with previously-amassed knowledge of how to earthquake-proof buildings) still remain intact. The natives that currently live nearby to the site say (just as others who live close by to other ancient sites say) that they have no knowledge of the ruins except for the fact they were built by the "gods".

These ancient sites that exist all over the world are grist to the mill for those who support the theory of "ancient astronauts". These archaeological sites divide opinion because we cannot understand how the ancients could have created them – so we assign them to the work either of a higher power (gods) or to more advanced beings who arrived on Earth as teachers of humanity. This latter is a valid opinion and perfectly possible. The alternative is that we have simply underestimated the abilities of the ancients. And there is a third question, if teachers taught us, then who taught the teachers?

Of course the "ooparts" (or, "out-of-place-artifacts") that were manufactured millions of years ago, stun us completely. They were constructed with very sophisticated ancient knowledge of electrical and mechanical engineering. Of course, when many of us begin to consider the idea that humans were around millions of years ago and busily producing clever artifacts, we baulk considerably. We prefer instead to fall back to the idea that these objects either fell to lower levels of the ground through natural movements within the earth, or that they were wrongly dated. For many of us, even to contemplate ancient skilled tilers – on their knees busily constructing tiled pavements more than 20 million years ago – is a step too far. This particular example was found in the Plateau Valley of Colorado when, in the 1930s, a man who was digging to create a storage space for winter vegetables, discovered a tiled pavement in geological strata that contained other animal artefacts that were believed to be 30 million years old. Some archaeologists, in the face of such discoveries, might be forgiven for simply walking away and making a quote such as Oliver Hardy's (when exasperated with Stanley) "I have nothing to say."

Just as astounding, is to imagine television in the time of Enoch the seventh patriarch. The Book of Enoch was removed from the Old Testament – likely because its content was not considered appropriate for inclusion (which was not unusual). Down through the ages, various parts of the bible have been interfered with, altered or removed if there was a chance that they might offend the current emperor, pope or king. It is widely accepted that the Roman Constantine was the first Christian emperor. Although he was pagan at heart, he was a pragmatic man. Christians were rapidly increasing in number and Constantine was struggling with a failing empire – so he accepted rather than persecuted them. He wanted their support and what better way to get it than to become a Christian himself – and if he still held allegiance to his old gods, then what difference would one more make? However, he did make sure that their guidebook or almanac – their Bible – was suitably modified to suit him and the Council of Nicaea. So he, along with a vetting group, got together and removed whatever they chose.

To return to the Book of Enoch. In his book, Enoch states that the "Azazel" (a group of "fallen angels") taught men to make "magic mirrors" in which distant scenes and people could clearly be seen. This sounds very much like a forerunner of television. The Greek Lucian of Samosata also wrote about a "magic mirror". In his *Vera Historia* (written

Ancient Science

in the second century A.D.), he writes that this "magic mirror" is a "looking glass of huge dimensions [no doubt he was comparing it to a hand-held mirror] and through it one hears everything that is said upon the Earth and by looking in it, sees all the cities and nations of the world." Was he watching The World Today and running through all the channels?! We have to ask what would motivate anyone to make up such a story.

As we saw earlier in this book, ancient India has been the source of many mysteries. The Hindu epics in particular are full of such stories. As we have covered, some of these stories appear to indicate that ancient India had, through advanced technology, produced the kind of weapons that threatened the annihilation of much of humankind. And we have to wonder, if these weapons were indeed used (and as the widespread physical evidence seems to suggest), how much humankind has evolved since then in its proclivity for self-destruction. It seems that humankind, despite its advanced intelligence and reasoning powers, has not moved a single step forward in its realisation that war is a complete abdication of human rational behaviour. We still constantly, and at all times on some area of the globe, pursue this activity. When scientists first realised the destructive potential of harnessing such horror, this was the point at which all enlightened peoples of the globe should have come together and arranged for all the written knowledge on how to execute this destructive power to be destroyed.

Yet what happens today? Every new invention is assessed for its potential in war and destruction. So what is this strange proclivity toward killing, war and destruction that has halted human advancement? Perhaps it is a fault or genetic disorder within the group of genes that govern human intelligence and rationality. Already in around the year 400 B.C., Plato had the wisdom to say "Only the dead have seen the end of war." So why does this situation prevail nearly 2500 years later? There are only two entities we can blame – either a divine creator (who, if exists, one would have thought would have been infallible) or an extra-terrestrial source the like of which has been popularised in the music, movies and games of the last one hundred years. Perhaps such beings, in their attempt to enhance the intellect of the human gene pool with their own genes, missed certain eventualities (not being of this Earth). And perhaps as a result of this, certain unforeseen results – certain problems in the gene pool and in human behaviour – were unwittingly created. And so perhaps, because of these mistakes, masses of proto-humans had to be destroyed (and ancient knowledge and civilisations lost with it). Of course, the question of who carried out this cull would then arise, and this would have to be researched. In any case, the cull was evidently not a complete success, once again making us wonder whether the creator (whether divine or physical) was fallible or infallible.

And evidently, not all the ancient wisdom was lost and some of it managed to be passed down through the ages. Around the turn of the twentieth century, the celebrated Russian writer known as Maxim Gorky met a Hindu, in the Caucasus, who was something of a magician. Gorky had an amazing experience with this mystic. The mystic asked him if he wanted to see something in an album that he was holding, and Gorky, thinking he meant conventional pictures of India, happily agreed. The Hindu magician then put the album on the writer's knee and asked him to turn the pages. Inside, were polished copper plates depicting beautiful Indian cities, temples, landscapes, and vistas, and he thoroughly enjoyed looking at them. When he had finished looking at the pictures, he returned the album to the Hindu. The yogi then blew on it and smiling said, "Now will you have another look?" In his writings Gorky said, "I opened the album and found nothing but blank copper plates, without a trace of any pictures." Gorky later mused, "remarkable people these Hindus." One might say that this "trick" could be duplicated by any of our prominent modern magicians. But we should also remember that the Hindu people of 4500 years ago were certainly remarkable if it is true that they constructed all those flying aircraft, vimanas and nuclear bombs.

Ancient Science

In his book, *Readable Relativity*, the British mathematician Clement Vavasor Durrell writes, "all events, past, present and future are present in our four-dimensional space-time continuum, a universe without past or present." If this is the case, then perhaps it is possible to tap into such time periods and any significant events that took place within them – rather like winding back a cassette tape or scene on a DVD.

The ancient Gnostic text, the *Vision of Isaiah* (dating to the second or third century A.D.), tells an interesting story. In it, the prophet Isaiah was taken up to Heaven where he saw God in eternity. Then, the "angel" who had taken him to paradise finally said that it was time to leave and go back to Earth. Isaiah then replied "Why so soon? I've been here only two hours [did Isaiah have a watch? Or are there clocks in Heaven?!]." The angel then replied, "not two hours but thirty-two years". The Prophet was naturally shocked by this but the angel comforted him by telling him that he would not have aged on his return – although the Earth, and the people he knew there, would have. In the twenty-first century, science tells us that in a photon or antimatter-drive space vehicle travelling at a velocity that approaches the speed of light, our astronauts would experience an identical shrinkage of time, virtually jumping into the future.

Stephen Spielberg utilised the idea of this space technology in a scene from his film Close Encounters of the Third Kind. In the film, the abducted flight crew of the infamous Flight 19 returned as young as they were when previously abducted a couple of decades earlier. This technology could also theoretically be utilised in the theory of biogenetic creation; the extra-terrestrial "creators", during their operations, may not have needed to experiment over thousands of years of human development. Instead, this technology would have enabled them to carry out subsequent stages at roughly the same time. And these individuals in any case, probably lived for a couple of centuries – just as the Old Testament tells us the ancient patriarchs did.

CHAPTER IX

NOAH AND THE WORLD SURVEYORS

In his book, *Secrets of the Lost Races*, Rene Noorbergen introduced an interesting account of the "survivors that emerged from the Ark of Noah on the Mountain of Ararat" and wrote that they eventually, headed by Noah himself, went "down into a post-diluvian world, which would be vastly changed from its original form." However, there must have been survivors of the great flood – there always are after catastrophic events. And as we mentioned earlier, the Egyptian priests made it clear to visiting Greeks and scholars that there have been, and that there will be again, great conflagrations that wipe out much of humanity and that those that survive would have to start all over again and learn as children.

Certainly, this was the purpose of the building of Noah's Ark; and Noah and his team, which of course consisted mostly of his family, would have fulfilled the role of helping humanity to start again. The idea that the great flood was indeed a great catastrophic global event, rather than a local one, is gaining in popularity. It is probably understandable that even if it was a relatively local event, the ancients (who probably did little exploring themselves) likely viewed the catastrophe of the lands they knew as being a "worldwide" phenomenon. But in fact, the sciences of geology and archaeology are finding more and more evidence of it indeed being a global event.

To accept this does not necessarily require a belief in the Old Testament idea that the flood was caused by the hand of a vengeful God in retribution for human wickedness. But whatever the cause, it is interesting to explore what happened when, after the flood, Noah and the other survivors stepped down off the Ark onto dry ground. They would likely first have acclimatised themselves to their new area before they descended into the chaos (that the flood had caused) below, and before beginning the mission that they were spared for, that is, the regeneration of the Earth. The place where they descended is known as the "apobaterion", or "landing place", and that place was at one time thought to be Mount Judi while more recently it is thought to have been Mount Ararat, in Turkey.

Rene Noorbergen certainly believes in a worldwide flood, and in the Biblical Noah and his self-built survival vessel, and he participated in expeditions up Mount Ararat in the hope of finding it, or at least finding its remnants. In *Strange Realities*, I explored the life of Noah and the actions that Noah and his team took after disembarking – in particular, the way in which the animals were carefully released in a particular order. They were evidently in no hurry to go among the carnage on the lower plains because they first, apparently, planted a vineyard. After they released the animals, the carnivores and various birds would have assisted in devouring the enormous amount of carrion, both human and animal, that must have been lying around.

One would have thought that the flood would have left what we call an archaeological "signature" of some kind – perhaps a signature of bone material. A geological signature was left, for example, after the catastrophe that wiped out the dinosaurs. In that particular case, the obliteration clearly indicated that a celestial event was responsible. This is evidenced in the geological signature of black ash, and more importantly, the thin layer of the element iridium (that comes with meteor strikes) that can be found all over the globe and that is dated to some sixty-five million years ago.

One thing is for certain, when the survivors encountered the patriarch Noah and his group, the latter would have seemed like gods from somewhere other than earth – and this is perhaps how such legends of gods began. Before this encounter, the survivors would have needed a long period of time to pull themselves together from a semi-primitive state after they had lost everything. In the generation or so when Noah and his group established themselves and still resided on the mountain near the grounded ship, the young earthly offspring of the survivors would likely have heard stories as to how their parents and grandparents survived.

Yet, they would not have needed to start completely from scratch – that is, by wearing skins and using bone clubs (assuming humans where ever like that). Instead, these antediluvian survivors would begin again in what could be called the "metallic age" (rather than earlier ages) – and they would have been aware of what they had to relearn. It just took Noah and his group to arrive and give them their instructions and guidance.

It is relevant to note again here that the great factory complex consisting of two hundred furnaces at Medzamor, in Armenia, it is only around 14 miles from Ararat. So this bronze-age complex would almost certainly have been known about by the learned ones from the pre-flood times. Perhaps these factories were even created by them when they recognised the features in the soil that indicated a good place to begin ore extraction. These amazing factories were mentioned in Science et Vie where a journalist named Jean Vidal expressed the belief that "These finds point to an unknown period of technological development." He went further to suggest that the complex was founded by the wise men of earlier civilisations who possessed knowledge that they had acquired during an age unknown to us (the antediluvian age) – and that this age deserves to be recognised as scientific and industrial. In the centuries during which the long-living pre-flood people existed, and after they had fanned out far and wide from the mountains of Ararat, they would have had plenty of time to regenerate technology where it was possible, and to educate the newer generations in all things that previously existed.

There is evidence that the learned flood survivors influenced, if not actually reached, all places on Earth. Certainly, Noah's great experience in boat building would have enabled them to make any sea journey no matter how far. Enormous mapping operations took place – possibly reaching as far as the Arctic and Antarctic regions because some the extremely ancient maps showed coves and inlets that are now buried

Noah and the World Surveyors

under ice. Even Alexander the Great apparently owned maps of "great antiquity".

In Guatemala, as in no doubt other places, the ancient Mayan people busily recorded the exploits of the learned ones. In their sacred book, the *Popol Vuh*, they recorded that the "first men" possessed great knowledge, that they knew all, and that they examined the four corners and the four points of the arch of the sky (this could only refer to the four points of the compass) and the round face of the Earth.

Many fragments of technological know-how of unknown origin have surfaced over the years and have usually been ignored by science because of the difficulties that they pose in finding logical explanations for them. The ooparts are a primary example.

Eventually, after Noah's group had trekked through the lands teaching and enlightening, they would have reached their ancient homeland and entered into the legends and stories that Solomon and Assurbanipal read about. Perhaps they even reached Tibet and remained there (or at least part of their number). Perhaps it was they who, through the generations, became the "ones that knew everything" that Apollonius of Tyana eventually encountered.

The descendants of the earthly survivors of the flood would have learned much from their parents through oral tradition. Natural human intelligence is a unique gift and will always enable humans as a species, in spite of anything thrown at them, to survive. But what of the idea that the flood was created to purge humankind of its "wickedness"? If we consider, for a moment, that there is any truth in this part of the story, then do we accept the idea that the "destroyers" who created the flood were willing to sacrifice any good people that existed along with the bad? And in turn, how many humans who possessed the unfortunate trait of "wickedness" survived the flood (perhaps through sheer luck)? It would not seem likely that in such an event all the survivors were as wise and as evolved as Noah, but perhaps, having escaped such a catastrophe, these individuals would have become repentant and god-fearing in their appreciation.

In spite of their alleged wickedness, the antediluvian people were highly knowledgeable in agriculture, animal husbandry, construction, architecture, political organisation, metalworking, music, instruments, abstract arts, mathematics, chronology and astronomy. Yet for all this, the survivors, who were the parents of the new generation, would have been too occupied in the everyday struggle to survive to have much time for coaching and instructing the new generation. So the new generation would have had to have learned through observation – for example by observing the building of survival units and learning how to grow food.

And so, the new generation would likely have welcomed their first lessons from the experts who had come from the Ark. We must assume that since we are talking about so many thousands of years ago, that all the great edifices such as the pyramids and so forth still had to be constructed – so a great amount of teaching and learning must have taken place in the attempt to bring the world back to its former glory.

But one thing that seems to have been lost was the longevity of the bulk of the population. Before the flood, when the Earth was in its idyllic state, they lived for centuries, but when the survivors had to work to rebuild the world in conditions that we can only imagine, perhaps this is what caused them to begin living shorter lives. Technical society is dependent on a large population for its maintenance – and during the time after the flood, not only was the surviving population very small, but the technical resources were gone, as was the availability of natural resources and the specialisation and coordination of labour. As a consequence, the degree of civilisation that the small surviving family from the Ark was able to re-establish was severely reduced.

But Noah and his family at least had longevity on their side. According to Genesis, Noah died at the age of 950 – which meant that he would have been born around the year 3998 B.C. and died around 3048 B.C. He had three sons named Shem, Ham and Japheth, and we have to

Noah and the World Surveyors

assume that they were also blessed with the genes of longevity. Of course, Noah and his family were capable of creating and teaching only those things that they had personal knowledge of; although they were in possession of a great amount of knowledge and although they would have had the memory of a technological environment to fall back on, they wouldn't necessarily have been skilled in all things. It is likely that the antediluvians (at least in part because of their longevity) used a far great portion of their brains than we do today. Their memory would have been vastly superior which meant that there was less need for writing. Instead, their knowledge was passed by word of mouth and memorised in the brains of their recipients or pupils.

Unsurprisingly, Genesis does not say anything about writings from other cultures that mention an equivalent flood story, and an equivalent Noah character, but there are many. For example, in Chinese mythology there is a character named "Nüwa" who is said to have re-created humankind after a great flood. Surely the sound of this name is just about as close as you could get to the Judaic/Christian patriarch? Perhaps all of these stories point to one common original story that has been lost so far back in the mists of time that it even predates all of these myths. Or perhaps there were a number of Noah-like characters who saved their families during this great calamity. If the latter is the case, it leads to the question of where else their ships may have come to rest. From which other locations did the antediluvian people spread their knowledge? Perhaps the Chinese Nüwa landed on the peaks of the Himalayas and founded the hierarchy of wise ones "who knew everything" in Tibet. And perhaps they founded the monasteries and learning centres that existed all across East Asia.

In our own civilisation, science has historically advanced by means of the scientific method which involves testing, experimentation, and observation. Technology has then taken the discoveries made by science and applied them into usable form. It must have been very frustrating for the patriarchs to have had to regenerate this process. The common idea that "history repeats itself" seem even more pertinent in this situation – especially with the knowledge that there may have been continual conflagrations, and resultant rises and falls of humanity, even before the flood into the mists of time. Will humanity ever be able to break this cycle? Perhaps some unknown entity or power in the cosmos ensures that human progress only ever goes so far and no further.

In *Secrets of the Lost Races*, Rene Noorbergen writes, "archaeology tells us nothing at all about the pre-deluge period, for most archaeologists completely disregard an antediluvian era in history, as they mostly theorise that the flood was merely a local event and not worth considering in the scheme of world history." When Noorbergen wrote these words around fifty years ago, this may have been true. But since then, opinion has changed somewhat, as seen, for example, in the fact that there have been expeditions to Mount Ararat in order to search for any traces or remains of the Ark of Noah. Regardless of criticism of the bible, there is a serious trend among archaeologist to explore the historical aspect of its stories.

And after all, the Bible is the best source we have for historical clues with regard to Noah and the patriarchs. In fact, it is quite specific in its detail. It is maybe precisely this detail that seems to discredit it for many people – for example when they read that the patriarchs lived to be centuries old, or when they read that they developed their advanced skills so soon after the birth of Adam the "first man" in 5000 B.C. (a fact that is evidenced in the age of the factories at Medzamor that date to the bronze age). The dating of these events is hard for many people to believe. Just as the story that Moses wrote in Genesis – of the world being created in seven days – also seems hard to take literally. The ecclesiastical hierarchy should really not be surprised that most scientists baulk at accepting these precise details.

Nevertheless, the survivors of the Ark were said to have moved out into the furthest lands, teaching and introducing new and more efficient methods of agriculture, teaching how to dig for special minerals and

Noah and the World Surveyors

metals, and how to undertake the smelting process and to make tools and (perhaps inadvertently) weapons. They taught geography, navigation, boat building and all the skills that were necessary for the post-diluvians to advance once again. And certainly, the world today contains a treasure of evidence pointing towards unceasing activity and advancement during this grey dawn of post-flood development. The scope of the navigational and surveying techniques utilised by these ancients should not be underestimated. For example, the patriarch Almodad, who was a descendant of Noah, was an intrepid explorer and geographer and was said to have measured the Earth "to its extremities."

The Indian nation, particularly the Hindus, learned very quickly during this time. The sacred Hindu books, known collectively as *The Puranas*, refer to direct communication between India and distant places around the world. Even Britain was known to them as "Sweta Saila" which translates to "the land of the white cliffs" in Sanskrit. Clearly, they had a great knowledge of geography.

We also have evidence of advanced geographical knowledge in ancient Chinese culture. One of the oldest surviving literary works from China is the *Shan Hai Jing*, or, the *Classic of Mountains and Seas*. This is a treatise on geography, the knowledge in which was obtained from a worldwide geographical survey that must have taken place soon after the flood. Other books have also emphasised the rich culture of exploration that existed in ancient culture: "all the prominent mountains of the world were known, plotted and depicted on either side of the oceans including the full length of the US, starting in Canada and down to South America. The Chinese themselves in ancient times were avid explorers and surveyors." So clearly, the middle-eastern patriarchs were not the only antediluvians, and not the only antediluvians to explore the world and leave their trace in the form of maps, symbols and place names.

So why was all this exploration undertaken in the antediluvian era? The most obvious explanation would be that, as Noah and his family stepped down from their survival vessel, the Ark, they would have looked down upon a world that was totally alien to them. All the familiar landmarks had disappeared: forests were gone, with rugged, bare mountain peaks standing in their place. And from the murky waters below rose the foul stench of decay. The Earth that they had once known was wiped completely clean of any previous civilisation; it was as if they had landed on another planet. So as the new generations were born and grew up on the foothills of Mount Ararat, their innate curiosity, and their desire to fulfil the ancestral promise of regenerating humankind and the world, would have encouraged them to venture out into the vast hinterland.

Because of the ancient standing stones that appear all over the world, and the way that these stones are positioned, we know that the antediluvians were aware of the earth's natural forces and what we call today the earth's "telluric currents". The practice of erecting these stones was regenerated during the rise of Egyptian culture when they placed numerous stone "needles" or pillars at certain points, and sometimes in patterns spreading outward. After they left the Ark, re-establishing the locations of these energy points, and marking them with stone containing high amounts of crystal, would have been quite high on their priority list. It is said that in pagan Britain, the witches, warlocks, "cunning men", and of course the Druids, were all quite familiar with these energy points. The church also knew of them and was evidently concerned that the power they held presented a threat to the supremacy of their Christian God and to their church.

However, the church had to tread softly; they knew that it was more advantageous to keep the people on their side rather than to alienate them by depriving them of their ancient beliefs and customs. The church knew that these mysterious people who dabbled in the paranormal (as we would call it today), provided a service of sorts to the masses. For example, a person might have gone to the "cunning man" or witch, well-versed in the subject of Wicca, for cures, charms and spells to help with an illnesses, a failed crop, or even to find a lost pet. Of course, there was also a dark side to these practices in that individuals might have wanted to obtain spells that wished harm or revenge on their enemies or neighbours. These darker spells often involved obtaining personal items, such as a piece of clothing or jewellery, from the object of their hatred, and including this in the ritual of the spell.

So while the church tolerated these practices to a certain point, there came a time in the fifteenth and sixteenth centuries when the church, and society more generally, began seeing witches everywhere. This obsession grew into such fear and hysteria that some individuals being accused of witchcraft were arrested solely on the evidence given by children. Enter Mathew Hopkins (c.1620–1647), the self-proclaimed "Witchfinder General" of England. Masquerading under the guise or righteousness, his goal was to persecute and murder all those suspected of partaking in these black arts. Other areas of authority did things in a more gradual manner – examples of which include the church superimposing Christian feast and saints' days onto ancient holidays, and building churches on geographical sites that were considered sacred in old pagan and Druid traditions.

Eventually, in the case of the latter, nearly all knowledge of the importance of these sacred points and earthly forces was lost. For many centuries, knowledge of them continued only among very few people. Today however, knowledge of sacred sites and the possible significance of phenomena such as "ley lines" is becoming more widespread again. And it is now also becoming clear that many old churches were deliberately constructed on these lines and points (more on this later in this chapter).

So, the early generations that emerged from the Ark moved out over the Earth instigating what could be called the first "land grab" in history – staking out their territories, settling, and creating nations for their children and the generations to come. And it is likely that this is the time at which many ancient archaeological sites and megaliths were built. The American writer and professor, Charles H Hapgood, has written about the placement of many of these sites; he pointed out that their positioning was often so precise and accurately-aligned with the cosmos, that the knowledge and skill of these early builders far surpassed anything we previously thought they had been capable of. And so it became more and more clear just how thorough the mapping and surveying of the "New Earth" had been. These people must have been in possession of accurate surveying instruments – and so it is likely that these important instruments had been saved with them on the Ark. Perhaps this surveying/mapping is the true meaning of what is referred to in the Bible when it is said that it was during the days of Peleg (one of Noah's descendants) that the Earth was "divided". His name is also associated with navigation and sailing.

This exploration, surveying, and mapping was carried out on a worldwide scale. And significantly, as the work of Charles Hapgood has shown, the maps show that they were originally created at a time when there was no ice on the Antarctic coastal land areas (which could help us to date the surveying of this coast to at least 4,000 B.C. if not before). One of the patriarchs who likely participated in this charting of the world was a grandson of Noah whose name was Mizraim whose name means to "delineate", or, to draw up a measured plan. His particular area of concern was the area now called Egypt, and some sources state that he was the founding father of the Egyptian people. The significance of Mizraim and of Egypt in the ancient world is hinted at by the fact that at least two of the most ancient maps were based on a circular projection with the central focal point being Egypt. Another descendant of Noah who most likely played a part in geographical surveying was Almodad, whose name in Hebrew means "measurer".

In the Chaldee paraphrase of Jonathan ben Uzziel, there is preserved an ancient tradition which tells us that Almodad was the inventor of geometry and the individual who "measured the Earth to its extremities." Highly intelligent and renowned Greeks, such as Pythagoras, are referred to as those ancients who were responsible for giving us so much but whose knowledge came from more ancient sources. And so we remember

Noah and the World Surveyors

again Solomon's phrase "there is nothing new under the Sun", and Assurbanipal, who got all his learning from reading about the accomplishments of the ancients, and the seventeenth-century enlightenment figure, Sir Isaac Newton, who said, "if I have seen further, it is because I have stood on the shoulders of giants." And so the chain goes back to the learned characters of ancient Greece who themselves stood on the shoulders of giants who existed even further back in time.

According to the records and details in Genesis regarding the professions and the ages of the patriarchs, it is evident that their lifespans overlapped. So this means that they would have worked together (or at least corresponded and collaborated) in this vast enterprise of re-mapping and re-settling the globe. And the main players in this endeavour were likely to have Peleg, Almodad and Mizraim.

When William and Mae Marie Coxon made a ten-year study of ancient signs and petroglyphs found all over the world, their conclusion was that at a very remote point in human history, a group of ancients they called "The Stone Writers" left their trace in the form of signs that can be found on every continent. By careful consideration, the Coxons discovered special sequences of particular geometric signs and symbols in different regions. And by dating the petroglyphic remains in the Nile Valley and comparing these with the later Egyptian civilisation, the Coxon's estimated the date of the stone writers' appearance as being about 1500 years before the rise of Egypt.

The repetition and location of the symbols that the Coxons found indicated that they had meaning and purpose. The Coxons deduced that they not only travelled the oceans but also penetrated far into the continents because they left their guide-marks along all the rivers for those who would follow them. They concluded that the stone writers were the explorers and geographers of old who "charted the world after the flood."

This work on the symbols and their meanings is verified by the work of other archaeologists such as the aforementioned Hapgood who also concluded that a system of signs was used by navigators many thousands of years ago. Others who have studied these symbols, as well as the origin of certain words across different languages, have concluded that the dispersal of the human race occurred from a common point of origin that is situated somewhere in the Middle East. And this idea fully corroborates the historical Genesis record and its story of the dispersal of nations from one point.

Professor Hapgood suggests that the mapping of a continent on such a vast scale indicates that an economic gain from the land must have been envisaged. So the expeditions did more than just discover and cultivate new areas for settling, they actually deliberately divided the earth into parcels of land (or nations) so that economic resources could be owned, shared, and traded. And, it has been shown that the dividing lines of these lands ran along what we now know as "ley lines"

This brings us back to the significance of churches being built upon sites that were once considered sacred to the pagans. One afternoon in the 1920's, a merchant named Alfred Watkins whose hobby was prehistory, was riding on horseback through the Bredwardine Hills near Hereford, England. On reaching the summit of a grassy hillock, he rested, letting his gaze fall over the tranquil English landscape. Suddenly, he saw something that he had never noticed before – several church steeples running in straight lines across the countryside. Knowing that these churches had been constructed on ancient sacred sites, he wondered whether it was possible that they had once been linked (or still were) by an unseen web of lines. While he pondered this question, he soon realised that not only were churches running in lines, but that archaeological mounds, standing stones, crosses, sacred trees, cross roads and moats with allegedly sacred wells, were also.

Hurrying home, Watkins painstakingly marked off all the ancient sites and monuments, that he knew of from his studies of local pre-history, onto a one-inch ordnance survey map. Even finding five or six points in alignment would have been beyond mere chance, but he found himself confronted with eight, nine and more, all stretching out in straight lines.

Carrying his initial research a step further, he compared his points to positions on other maps he had marked and discovered that the lines could be extended for many miles, usually ending up at a mountain peak, or a high cliff. So, aided by a friend, Watkins undertook a detailed survey of the whole of England and Scotland; and throughout the whole area they found further traces of a prehistoric network of sacred sites in straight alignments extending over the entire island of Britain.

Building on the accomplishments of Watkins, Major H Taylor of the British Army set out, with the help of a professional surveyor, to undertake an even more detailed study of these mysterious lines. Taylor discovered more landmarks that were previously unknown (or at least not recorded) in modern times. His findings were eventually published in a short book titled *The Geometric Arrangement of Ancient Sites* in 1939.

But these lines had not only been noticed in Britain. A year prior to the publishing of Taylor's book, a German geographer, Joseph Heinsch, presented a paper at the Geographical Congress in Amsterdam, that detailed the same type of discoveries. His paper was titled *Grundsätze vorzeitlicher Kultgeographie*, or, *Principles of Ancient Cult Geography* and in it, he proposed to a hushed audience that in distant times past, sacred sites had been situated in the landscape according to a set of "magical" principles. He proposed that these sacred sites were constructed on lines, and that the pattern of these sites and lines in the landscape, related to the positions of the Sun, Moon and planets. Furthermore, he claimed that he had uncovered evidence that the units of measurement evident in the pattern of these lines were like those of the early Egyptian geodetic surveys that were based on simple fractions of the Earth's dimensions. He had found examples of these lines not only in Britain but also all over Europe and the Middle East.

Greatly impressed by the vast extent and accuracy of the construction along these lines, Heinsch concluded that they bore testimony to the existence of a widespread ancient civilisation that possessed advanced knowledge of both technology in the form of accurate instruments, and not a little magic. Today, these "magic" routes have become well-worn roads and footpaths – and as Heinsch showed, they stretch far beyond the boundaries of Britain and to every corner of the globe. But what are these lines? It has been shown that they possess a kind of "magic" energy that might perhaps be described as a "telluric force". In his book *The Fairy Faith of Celtic Countries*, W. Y. Evans-Wentz, recalls how an old Irish seer explained to him that mysterious currents flow along paths but that their exact nature has been forgotten.

Similar research conducted by the French archaeologist and writer, Xavier Guichard, strongly supported the findings of the British and German investigators in relation to old cities in France. Guichard wrote, "These cities were established in very ancient times according to immutable astronomical lines, determined first in the sky, then transferred to the Earth at regular intervals, each equal to a three hundred and sixtieth part of the globe [that is, every degree of the globe]."

Evidence that knowledge of these lines existed in remote history can also be found in ancient literature. For example, in their writings about their conquest of the Etruscans, the early Romans noted the standing stones that were set in linear patterns over the entire countryside of Tuscany. Later, during the Roman invasion of Greece, they recounted the fact that stone pillars were found running straight and true along the roads of the hilly Hellenic landscape. So the Romans were not particularly surprised at finding these straight tracks because they had discovered them in almost every country that they conquered – from across Europe, North Africa, Crete and as far west as the regions of ancient Babylon and Nineveh. We now recognise that the Romans' reputation as builders of straight thoroughfares was partly attributed to them simply utilising the sacred lines that already existed long before their conquests, and which they then transformed into military and commercial routes.

Noah and the World Surveyors

Even in areas where there are no roads, navigation signs exist. Leading up to fairly recent times, the Bedouin of North Africa used a system of lines that were marked out by standing stones and cairns to help them traverse the desert wastes. But even though these stones are essential markers of the Bedouins' survival, their origin and date has been lost in the mists of time. So it seems that these markers and sites that traverse the globe are a tribute to the determination, energy and indeed accuracy of the original surveyors who left a legacy that is still apparent today.

A big part of their work would have been having the necessary ocean-going skills to travel as far as they did. There was probably no better a boat-builder than their predecessor Noah, whose skill in boat building made their endeavours possible. They evidently travelled across the Pacific Ocean, as the ancient stories of the Australian Aborigines testify. The Aborigines tell of a past age that they call "dream time" when the "creative gods" traversed the country and reshaped the land in order to conform to the important paths they call "Turingas". And Aboriginal tradition says that at certain times of the year, the Turingas become revitalised by the energies flowing through them – and that this gives life to the adjacent countryside. Still today, to ensure that this ancient fertilisation takes place, the Australian Aborigines gather at specific locations at certain times of the year, to perform rituals and dances (that have been dictated through time) and to pray to the "force" of the lands.

Knowledge of this type can be found in China also; for example, in the Chinese practice of Feng Shui (sometimes called "Chinese geomancy") these forces were known as "dragon currents". Stone formations have also been found in China, but in all places, the placement of stones appears to relate to earth currents in some way. It was also likely that the ancients were aware of the mineral content of different stones because stones with a high crystal content were often used. In Patterns of the Past, Guy Underwood suggests that standing stones upon the Earth served the same purpose as reinvigorating the human body with needles in Chinese acupuncture – both of these practices have, and had, the aim of restoring balance and vitality.

All this is not simple mythology; scientific data to support the existence of these strange forces is increasingly emerging. For example, it is known that the entire Earth's surface is bathed in the Earth's magnetic field and the energy of the Sun. And, the strength and the direction of the magnetic currents vary according to the positions of the Sun, the Moon and even the closer planets The latter fact provides potential evidence for the argument that astrology is more than just guesswork. It is known that the fluids of the human body and brain are sensitive to these cosmic forces; and after all, considering that the human body consists of at least 70% water, and that the Moon influences the movement of the tides of the vast oceans, then the idea that the Moon has the potential to influence human bodies doesn't seem far-fetched. The word "lunatic" comes from the Latin word for "moon" – another clue that the ancient peoples were aware of the Moon's ability to affect the brain. With a full Moon being strongly charged with photons from the Sun, it is this that might cause the fluids and neurons in the human body and brain to become affected. Animals are also affected – for example we have the idea that wolves and coyotes howl more during a full moon.

After exhaustive research involving over 200,000 experiments over ten years, Giorgio Piccardi (1895–1972), director of the Institute of Physical Chemistry in Florence, Italy, concluded that water is extremely sensitive to magnetic fields. He discovered that when water is exposed to a magnetic field, it can cause the chemistry of the water to be altered. Piccardi also explored (and found to be true) the idea that the Earth's magnetic field is subject to change, and is affected by changes in the positions of the Sun and Moon. Piccardi's work has been verified by W. H. Fisher, of the National Centre for Atmospheric Research, Boulder, Colorado, who argued that since water is the liquid of life, this means that potentially, electromagnetic fluctuations could affect humans and human health.

Horticulturists have also noted that electromagnetic fluctuations can affect the growth of plants. Let us return, for a moment, to the wise ones or "cunning men" who were prevalent in medieval society before religious pressure caused them to go underground. These are the individuals to whom villagers would have gone for help, for example, if their corn was not flourishing or their tomatoes not ripening. It is almost certain that these wise individuals were knowledgeable of, and aware of how to work with, the Earth's forces. And the effect of magnetic and electrical forces on plant life is now also being verified by science. For example, two horticulturalists at Utah State University, A. A. Boe and D. K. Salunkhe, found that when green tomatoes were placed within a magnetic field, they ripened four to six time faster than under their usual conditions. And these studies concluded that magnetic fluctuations have the ability not only to affect plant fertility but also to affect the mineral content and chemistry of the soil. So while we are only just now beginning to understand the scientific principles behind the celestial and terrestrial magnetic forces, and their effect on the Earth and the creatures that walk upon it, the people of centuries past already knew about these forces and knew how to work with them for their own benefit.

The amazingly skilled explorers and mappers of antiquity (who we assume were the descendants of Noah) appear also to have been in possession of this knowledge – otherwise how, and why, would they have navigated and surveyed in the way that they did? And surely this type of knowledge would have been the end result of generations of accumulated experimentation and knowledge. So, it seems clear, that the primary mission of the patriarchs' descendants was to restore and reinvigorate the Earth in order to promote the re-growth and the health of the forests, plants and animals after the devastation of the flood. And it also seems likely that there was a centralised point somewhere that this knowledge radiated out from – rather than the more unlikely idea that all these simultaneous practices arose all around the world by pure coincidence. In *The View Over Atlantis,* John Mitchell wrote, "A great scientific instrument lies sprawled over the entire surface of the globe. At some period thousands of years ago, almost every corner of the Earth was visited by a group of men who came with a particular task to accomplish."

And so it seems that at some point, when the post-flood settlements around the globe were forming active cultures, that these peoples were trying to re-establish the "one world" civilisation that had existed during the antediluvian period. And we have clues that point to the existence of antediluvian civilization in the form of the amazing Ooparts that survive from time periods that are hard to conceive of. Thanks to these artifacts, we can begin to envisage a time period that existed even before the patriarchs and before 5000 B.C. (when the "first man" Adam was born) – a time period when an even more advanced technology existed.

Yet would we not naturally assume that any advanced technological society would go hand-in-hand with a society that displayed civilized human behaviour patterns? Apparently not. The entire flood saga revolves around the idea of the "wickedness" of men at that time. And this was not just confined to a couple of cities such as Sodom and Gomorrah, but apparently the entire globe. And if there is any truth in the story at all, we have to ask ourselves, who were the destroyers of most of the world's population? Surely such an action could not have been initiated by a kind or all-forgiving god? It sounds more like the decision of a cold and impassive intelligence that decided that its human experiment had failed and that it needed to wipe the slate clean before it could start again.

Genesis and all the older Judaic texts that contributed to the Bible, have lasted for thousands of years and will no doubt be studied and puzzled over for many years to come. We will continue to ponder questions such as how the birth date of Adam around 5,000 B.C. can be reconciled with the discovery of ooparts that were made a million years ago. Or questions such as, if scientists tell us that the Universe was created a long time before our solar system, then why does Moses tell us that creation was all over in a week? No doubt it was not for lack of education that these stories were written in this way, but rather, that they were written in a style that was deliberately symbolic, and that they drew upon stories that were already ancient at the time the Bible was written. The ancients that tutored Moses in the second millennium B.C. were the same ancients whose knowledge and culture fed into Greek civilization – the civilization who knew, for example, that life began in the sea, and who are now known for their great philosophers, physicists, mathematicians and astronomers.

But as we have discussed, cultures can lose their knowledge, civilizations evidently rise and fall, and as Graham Hancock says, "we are a species with amnesia." But it seems that now in the twenty-first century when new archaeological discoveries are coming to light all the time, we can no longer bury our heads in the sand. It is time to "remember" and blind faith is no longer acceptable – whether that be the blind faith of a scientific community who refuses to investigate new evidence, or the blind faith of a religious community who stick too rigidly to literal interpretations of their ancient texts.

The End

www.ingramcontent.com/pod-product-compliance
Lightning Source LLC
Chambersburg PA
CBHW020803160426
43192CB00006B/425